Matrix Ensembles in the Many-Nucleon Problem

Matrix Ensembles
in the
Many-Nucleon Problem

NAZAKAT ULLAH
Tata Institute of Fundamental Research, Bombay

CLARENDON PRESS · OXFORD
1987

Oxford University Press, Walton Street, Oxford OX2 6DP
Oxford New York Toronto
Delhi Bombay Calcutta Madras Karachi
Petaling Jaya Singapore Hong Kong Tokyo
Nairobi Dar es Salaam Cape Town
Melbourne Auckland

and associated companies in
Beirut Berlin Ibadan Nicosia

Oxford is a trade mark of Oxford University Press

Published in the United States
by Oxford University Press, New York

© Nazakat Ullah, 1987

All rights reserved. No part of this publication may be reproduced,
stored in a retrieval system, or transmitted, in any form or by any means,
electronic, mechanical, photocopying, recording, or otherwise, without
the prior permission of Oxford University Press

British Library Cataloguing in Publication Data
Ullah, Nazakat
Matrix ensembles in the many-nucleon
problem.—(Oxford studies in nuclear physics).
1. Nuclear structure
I. Title
539.7'4   QC793.3.S8
ISBN 0-19-851725-4

Library of Congress Cataloging in Publication Data
Nazakat Ullah
Matrix ensembles in the many-nucleon problem.
(Oxford studies in nuclear physics)
"TIFR/TH/85–9."
Includes bibliographies and index.
1. Nuclear structure—Statistical methods.
2. Nuclear reactions—Statistical methods.
3. Matrices. I. Title. II. Series.
QC793.3.S8N39 1987   539.7'4   87-19108
ISBN 0-19-851725-4

Photoset and printed at The Universities Press (Belfast) Ltd.

PREFACE

The subject of matrix ensembles has grown steadily in the last 30 years. In the early stages matrix ensembles were used primarily to analyse the vast amount of data on compound nucleus reactions. Matrix ensembles successfully explained the distribution of the widths and spacings of compound nucleus levels. Many theorists were attracted by the mathematical physics aspect of the subject and they derived rigorously many of the well-known results. As time passed, the matrix ensemble found many applications in the field of many-nucleon physics. In this book I will discuss applications to such topics as the collective parameters of nuclei, average cross-sections, and many other subjects in many-body physics.

The principal aim of this book is to assist young graduate students to understand the basic ideas and mathematical formulations of matrix ensembles. It is hoped that this book will also be useful to experimentalists who would like to use some of these techniques to analyse their data.

It is my pleasant duty to thank many of my colleagues and other physicists who have given their suggestions while this book was in preparation. In particular I would like to thank Drs S. K. Bhattacharjee, S. Mukhi, R. K. Nesbet, S. P. Pandya, J. C. Parikh, D. J. Rowe, V. Singh, C. S. Warke, and H. A. Weidenmüller.

I would like to dedicate this book to the memory of my grandfather.

Bombay   N. U.
January 1987

CONTENTS

1. Statistical aspects of the many-body problem   **1**
   1.1 Introduction   1
   1.2 Wigner's matrix ensemble   2
   1.3 The distribution of the Hamiltonian matrix elements   5
   1.4 New mathematical techniques for evaluating ensemble averages   6
   1.5 Applications of the random matrix ensembles to different areas of physics   7
   1.6 Concluding remarks   9
   1.7 References   9

2. Collective rotational parameters of nuclei   **11**
   2.1 Introduction   11
   2.2 Method of moments   11
   2.3 Numerical results   13
   2.4 Correction to the $J(J+1)$ spectrum   15
   2.5 Semi-classical formulation   16
   2.6 References   17

3. Vibrational nuclei   **18**
   3.1 Introduction   18
   3.2 Description in terms of an intrinsic state   19
   3.3 Derivation of an expression for ω 20
   3.4 Peierls–Yoccoz type of formulation for the vibrational parameters   21
   3.5 Model calculation   23
   3.6 References   24

4. Transition region in heavy nuclei   **25**
   4.1 Introduction   25
   4.2 Coherent phonon model   26
   4.3 Energy expression in the rotation–vibration region   27
   4.4 Energy expression based on Bohr–Mottelson's collective Hamiltonian   28
   4.5 Fitting of the experimental data   29
   4.6 Concluding remarks   30
   4.7 References   30

5. Symmetry mixing — 32
5.1 Introduction — 32
5.2 Derivation of the expression for the mixing parameter — 32
5.3 Examples of symmetry mixing — 34
5.4 References — 35

6. Anticorrelation in many-body systems — 36
6.1 Introduction — 36
6.2 The correlation coefficient and energetically good wave functions — 37
6.3 Anticorrelation and the exact wave function — 38
6.4 Examples of the use of anticorrelation identities — 40
6.5 Concluding remarks — 42
6.6 References — 42

7. Approximate evaluation of certain physical quantities — 43
7.1 Introduction — 43
7.2 Approximate evaluation based on Taylor series expansion — 43
7.3 Approximation based on diagonalization in truncated space — 44
7.4 Method based on the use of lower-order moments — 46
7.5 References — 49

8. Correlations between the parameters of the S-matrix — 50
8.1 Introduction — 50
8.2 The relation between the S- and R-matrices — 50
8.3 The statistical distribution of the R-matrix parameters — 52
8.4 Statistical properties of the parameters of the S-matrix — 57
8.5 Average value of the scattering matrix and the fluctuation of cross-sections — 63
8.6 Some general remarks and current developments — 69
8.7 References — 69

9. Highly convergent expansion of the Rayleigh–Schrödinger perturbation series — 71
9.1 Introduction — 71
9.2 Operator method — 71
9.3 Method based on the linearization technique — 73
9.4 Concluding remarks — 74
9.5 References — 75

10. Application of Grassman integration in matrix ensemble theory — **76**
- 10.1 Introduction — 76
- 10.2 Grassman integration — 76
- 10.3 Probability density of a single eigenvalue — 79
- 10.4 References — 90

11. Averages over the space of complex orthogonal matrices — **92**
- 11.1 Introduction — 92
- 11.2 Statistical collision matrix — 92
- 11.3 Averages of the parameters of the statistical collision matrix — 94
- 11.4 Average value and fluctuation of various cross-sections — 104
- 11.5 Concluding remarks — 112
- 11.6 References — 113

12. Configuration interaction problem — **114**
- 12.1 Introduction — 114
- 12.2 The configuration interaction problem using statistical methods — 115
- 12.3 Perturbative statistical method — 121
- 12.4 Probability density function of the lowest eigenvalue — 126
- 12.5 References — 132

Appendix A. Expressions for the first few orders of ordinary perturbation theory — **134**

Subject index — **135**

Author index — **139**

1
STATISTICAL ASPECTS OF THE MANY-BODY PROBLEM

1.1 Introduction

We are all familiar with the quantum mechanical description of the single-particle system, which considers the motion of a particle under the influence of a harmonic force. The task here is to find the eigenvalues and eigenfunctions of the single-particle Hamiltonian. Even in the single-particle case symmetries are exploited; for example, spherical symmetry in the case of the hydrogen atom makes it easier to find its possible energy levels. It is soon clear that exact analytic results can be obtained for only a few single-particle Hamiltonians. For most Hamiltonians variational principles and perturbation theory must be used. If this is true for a single-particle problem, it is obvious that life will be more complicated when the number of particles increases.

The first many-particle problem was the multi-electron system. It was not difficult to write the Hamiltonian for such a system but it took quite a while to derive the correct form of the many-electron wave function which, in the zero-order approximation, can be written as a Slater determinant. The single-electron wave functions were obtained by solving the Hartree–Fock equations in a consistent fashion. However, in order to obtain better energy eigenvalues for the many-electron Hamiltonian it is necessary to go beyond the zero-order approximation and to use perturbation theory which requires the use of a large number of electronic configurations. Thus we can see the need for some kind of statistical treatment of these large configurations.

This need increases when we consider a system containing a large number of nucleons. There are two reasons for this: (1) There are two kinds of particles, namely neutrons and protons, in the nucleus; thus the number of configurations is much larger; (2) Unlike the electronic case, the nuclear potential is not fully known. At present there are attempts to derive it using the quark model of nucleons.

Historically, the first quantity to be evaluated using statistical methods was the nuclear level density which, in principle, could be obtained by counting the number of levels for a given number of nucleons distributed in some given shell model configurations. However, this method is only suitable for low energies. The quantity which enters into the formulation of level densities is the state density which is defined as the density of eigenstates counting each M state separately, where M is the z-projection of $\mathbf{I}$, $\mathbf{I}$ being nuclear spin. It has been shown (Huizenga and Moretto 1972) that state density can be calculated by calculating the nuclear grand partition function and by realizing that the grand partition function is the Laplace transform of state density. Using the method of steepest descent to evaluate the Laplace transform and using a nuclear model in which single-particle levels are equidistant, we find that the total level density $\rho(E)$ can be written

$$\rho(E) = CE^{-\frac{5}{4}} \exp\{2(aE)^{\frac{1}{2}}\} \tag{1.1}$$

where C and a are constants. An explicit derivation of this result is given by Huizenga and Moretto (1972). This was the first application of the usual statistical mechanics to a many-nucleon system.

1.2 Wigner's matrix-ensembles

The kind of statistical averaging which found very wide applications in nuclear physics started with the measurement of level widths and the positions of resonances in Bohr's compound nucleus. Because of the complexity of the problem it was not possible to calculate the width of each individual resonance starting from some reasonable two-body nuclear Hamiltonian. It was Wigner's (1957) idea to replace the diagonalization by a probability distribution problem. He assumed that each matrix element of the Hamiltonian has an independent Gaussian distribution, the dispersion of the off-diagonal elements being half that of the diagonal ones. Thus, if $P(\{H_{ij}\})$ denotes the joint probability density of the Hamiltonian matrix elements H_{ij},

$$P(\{H_{ij}\}) = K \exp\left(-\frac{1}{2\sigma^2} \sum_{i,j} H_{ij}^2\right) \tag{1.2}$$

where σ^2 is the mean-square deviation of the diagonal elements and K is the normalization constant. It ensures that

$$\int P(\{H_{ij}\}) \prod_{i \leq j} dH_{ij} = 1. \tag{1.3}$$

In writing eqns (1.2) and (1.3) we assume that the Hamiltonian matrix is real symmetric. A detailed account of this distribution is given in Chapter 8.

We see that the problem of finding eigenvalues by diagonalizing Hamiltonian matrix elements is now replaced by the problem of finding the joint distribution of the eigenvalues starting from the distribution given by eqn (1.2).

1.2 WIGNER'S MATRIX-ENSEMBLES

Before we discuss the consequences of eqn (1.2), we would like to ask: what is the difference between the usual statistical mechanics ensemble and the one introduced by Wigner? The difference is this. In the usual statistical mechanics we average over the set of eigenfunctions of the same Hamiltonian while in Wigner's matrix ensembles we average over the set of Hamiltonians (Brody et al. 1981).

Let us now return to the consequences of the distribution (1.2). Two main theoretical results emerge. First, if we consider the distribution of the spacing S between neighbouring levels, its probability density function

$$P(S) = \frac{\pi S}{2D} \exp\left(-\frac{\pi S^2}{4D^2}\right) \qquad (1.4)$$

where D is the average level spacing. This is known as the 'Wigner surmise'. As can be seen, the probability for small spacing S is proportional to S, which shows that nearby levels repel each other.

The second result is that the distribution of the widths of the compound-nucleus levels

$$P(\Gamma_\lambda) = (2\pi)^{-\frac{1}{2}}(\langle\Gamma\rangle)^{-1}\left(\frac{\Gamma_\lambda}{\langle\Gamma\rangle}\right)^{-\frac{1}{2}} \exp\left(-\frac{\Gamma_\lambda}{2\langle\Gamma\rangle}\right) \qquad (1.5)$$

where Γ_λ denotes the width of level λ and $\langle\Gamma\rangle$ is the average value of Γ_λ. This is known as the Porter–Thomas distribution.

The distributions given by eqns (1.4) and (1.5) fit the data very well. As mentioned earlier, other consequences are described in detail in Chapter 8.

An extremely important result in the theory of matrix ensembles was the realization that the distribution of the eigenvalues is independent of the distribution of eigenvectors. If we write the matrix of H as

$$H = TE\tilde{T} \qquad (1.6)$$

where E is the diagonal matrix consisting of N eigenvalues and T is an orthogonal matrix, we can independently study the properties of the eigenfunctions X_λ of the Hamiltonian H. If ψ_μ denotes the orthonormal basis set, then

$$X_\lambda = \sum_\mu T_{\mu\lambda} \psi_\mu. \qquad (1.7)$$

The normalization of $|X_\lambda\rangle$ implies that

$$\sum_{\mu=1}^{N} T_{\mu\lambda}^2 = 1. \qquad (1.8)$$

Thus, the statistical average of any quantity associated with $|X_\lambda\rangle$ can be obtained by writing the probability density of $T_{\mu\lambda}$

$$P(\{T_{\mu\lambda}\}) = K\delta\left(\sum_{\mu=1}^{N} T_{\mu\lambda}^2 - 1\right) \quad (1.9)$$

where K is the normalization constant.

It follows from this that the central limit theorem of probability theory can also be used to derive, for example, the distribution of the widths of compound nucleus levels.

Thus we see how Wigner's use of matrix ensembles to study the eigenvalues and eigenfunctions of non-relativistic Hamiltonians can be applied to a vast number of problems in many-body physics. The present book is mainly devoted to such applications of matrix ensembles to the many-nucleon problem.

To illustrate the kinds of averages talked about in matrix ensemble theory we consider the expectation value of some operator Q in the eigenstate $|X_\lambda\rangle$. Using eqn (1.7) the expectation value

$$\langle X_\lambda| Q |X_\lambda\rangle = \sum_{\mu,\nu} T_{\mu\lambda} T_{\nu\lambda} \langle \psi_\mu| Q |\psi_\nu\rangle. \quad (1.10)$$

Taking ψ_μ as the basis set in which Q is diagonal and denoting $\langle \psi_\mu| Q |\psi_\mu\rangle$ by q_μ and $\langle X_\lambda| Q |X_\lambda\rangle$ by $Q_{\lambda\lambda}$, we can rewrite (1.10) as

$$Q_{\lambda\lambda} = \sum_\mu T_{\mu\lambda}^2 q_\mu. \quad (1.11)$$

If we now assume that the $T_{\mu\lambda}$ are distributed according to eqn (1.9), this implies that $Q_{\lambda\lambda}$ will itself have some probability distribution. For simplicity, we consider a two-dimensional case, in which

$$T_{11} = \cos\theta, \qquad T_{12} = \sin\theta \quad (1.12)$$

and θ is distributed uniformly for $0 \leq \theta \leq 2\pi$ or, in other words,

$$P(\theta) = \frac{1}{2\pi}, \qquad 0 \leq \theta \leq 2\pi. \quad (1.13)$$

Writing Q_{11} as Q, we can write its probability density function using eqns (1.11)–(1.13) as

$$P(Q) = \frac{1}{2}\int_0^{2\pi} d\theta \delta[Q - q_1^2 \cos^2\theta - q_2 \sin^2\theta] \, d\theta. \quad (1.14)$$

On integration, this gives

$$P(Q) = \frac{1}{\pi}\left(\frac{2}{q_1 - q_2}\right)^2 \left[\left(\frac{q_1 - q_2}{2}\right)^2 - [Q - \tfrac{1}{2}(q_1 + q_2)]^2\right]^{\frac{1}{2}}. \quad (1.15)$$

Thus we see that in two dimensions Q has large probability near $Q = q_1$ or q_2.

For large values of dimension N this result changes; in that case Q has large probability at the average value of Q, namely $((1/N)\sum_{\mu=1}^{N} q_\mu)$. This also agrees with the central limit theorem of probability.

After the initial success of the idea of the random matrix ensemble, three developments took place: (1) a more basic understanding of the distribution of the Hamiltonian matrix elements; (2) development of new mathematical techniques to evaluate various ensemble averages; and (3) further applications of these ideas to other problems in many-body physics.

1.3 The distribution of the Hamiltonian matrix elements

We recall that in writing the distribution (1.2) of Hamiltonian matrix elements we assume that the Hamiltonian matrix is real symmetric and has a Gaussian distribution. The question which we now ask is whether there is a more fundamental way of arriving at this result. The answer is yes. The fact that the Hamiltonian matrix can be taken to be real symmetric follows (Porter 1965) if we invoke the invariance of the Hamiltonian operator under time reversal and rotation. This is the situation for most physical systems. A very general analysis was carried out by Dyson (1962) who showed that Gaussian ensembles can be classified as orthogonal (GOE), unitary (GUE), or symplectic (GSE) depending upon how they behave under rotations and time reversal invariance. The ensemble described by the distribution given by (1.2) is therefore called a Gaussian Orthogonal Ensemble (GOE).

Balian (see Brody et al. 1981) showed that the Gaussian nature of the ensemble can be derived by information theory. He introduced the functional

$$I\{P(H)\} = \int dH P(H) \ln P(H) \tag{1.16}$$

where dH stands for $\prod_{i \leq j} dH_{ij}$ and $P(H)$ is the joint probability of the matrix elements H_{ij}. Now minimize $I\{P(H)\}$ subject to the constraint that $\langle \text{tr } H^2 \rangle$ is some given quantity. The bracket denotes the ensemble average. This gives

$$P(H) = K \exp \lambda (\text{tr } H^2) \tag{1.17}$$

where λ is the Lagrange multiplier, and the constant K is fixed by the normalization condition.

Dyson next introduced 'circular ensembles' (see Mehta 1967). The need for circular ensembles arose from an undesirable feature of the

Gaussian ensembles, namely that all the matrix elements are not equally weighted. The essential idea of circular ensembles is that the system is characterized by a unitary matrix with eigenvalues $\exp(i\theta_j)$; it is assumed that the consecutive levels of the actual system behave in the same way as the angles θ_j. An important result (Mehta 1971) is that, in the limit of large dimensions, both the Gaussian and circular ensembles have the same fluctuation properties. The mathematical aspects of circular ensembles were discussed in detail by Mehta (1967).

So far we have discussed matrix ensembles in which the Hamiltonian matrix elements are assumed to have some given kind of probability distribution. To put in the broader features of nuclear spectra French (see Brody et al. 1981; Wong 1986) suggested that only the two-body part of the nuclear Hamiltonian, rather than the N-body Hamiltonian, should be treated as random. These ensembles are called two-body random matrix ensembles. Because of these finer features it becomes quite difficult to work out analytic expressions, e.g. for the single-eigenvalue probability density function. The success of these ensembles is to explain the shell-model density of eigenvalues which is found to be Gaussian rather than semicircular in form. Wong (1986) described the application of these ensembles to different nuclear problems. Some of the new quantities introduced in these ensembles are found to be very useful in discussing symmetries in the many-nucleon system and in the perturbative treatment of the lowest eigenvalues. They are described in Chapters 5 and 12, respectively.

1.4 New mathematical techniques for evaluating ensemble averages

It is easy to show from eqn (1.2) that, if the E_λs are the eigenvalues of the real symmetric Hamiltonian H, their joint probability density function $P(\{E_\lambda\})$ is given (Metha 1967; Porter 1965) by

$$P(\{E_\lambda\}) = K \exp\left(-\frac{1}{2\sigma^2} \sum_{\lambda=1}^{N} E_\lambda^2\right) \prod_{\mu<\nu} |E_\mu - E_\nu|. \quad (1.18)$$

Suppose we want to obtain the probability density function of a single eigenvalue, say E_1,

$$P(E_1) = \int P(\{E_\lambda\}) \prod_{\lambda \neq 1} dE_\lambda. \quad (1.19)$$

It is easy to see from eqns (1.18) and (1.19) that, because of the absolute value sign in eqn (1.18), this integration is very difficult to carry out except when the dimension N is small. Thus new techniques are needed to carry out further integrations using eqn (1.18). The technique developed by Mehta (1967) is called integration over alternate variables.

In this technique eqn (1.18) is first rewritten as a determinant

$$P(\{E_\lambda\}) = K |\Phi| \qquad (1.19)$$

where Φ is an antisymmetric normalized $N \times N$ determinant made up of single-particle orbitals $\phi_i(E_k)$ given by

$$\phi_i(E_k) = (\pi^{\frac{1}{2}}\sigma 2^i i!)^{-\frac{1}{2}} H_i\left(\frac{E_k}{\sigma}\right) \exp\left(-\frac{E_k^2}{2\sigma^2}\right) \qquad (1.20)$$

where $H_i(E_k/\sigma)$ are Hermite polynomials.

To get rid of the absolute value sign in eqn (1.19), we integrate over a region $-\infty < E_1 \leq E_2 \leq \ldots \leq E_N < \infty$ in which Φ remains positive. For further details refer to Mehta (1967) where expressions for the normalization constant K and other ensemble averages of the eigenvalues E_λ are given.

We now discuss the averages over the eigenvector components $T_{\mu\lambda}$. Because of the orthonormality condition

$$\sum_\mu T_{\mu\lambda} T_{\mu\lambda'} = \delta_{\lambda\lambda'}, \qquad (1.21)$$

we introduce Dirac delta functions to take care of these constraints. Further details of integrations using δ-functions are given in Chapter 8 where we discuss correlations of the parameters of the low-energy scattering matrix S. We remark that the probability density function of many quantities can be worked out using δ-function technique; a simple example is the expectation value $Q_{\lambda\lambda}$ given by eqn (1.11).

In recent times (Verbarrschot et al. 1984), Grassmann integration has been found very useful in studying the average resolvent. This is discussed in Chapter 10.

1.5 Applications of the random matrix ensembles to different areas of physics

As mentioned earlier the random matrix ensembles were introduced to describe the average parameters of the compound nucleus. If the energy of the bombarding neutron beam is low, isolated compound nucleus resonances are observed. From the observed data, say, of the widths of the levels, histograms (Kendall 1945) are constructed. The smooth curve which fits such histograms is the Porter–Thomas distribution (1.5). As the energy of the neutron beam is increased, the resonances start to interfere. Much theoretical research in the area of compound nucleus reactions is devoted to deriving expressions for the average reaction cross-section, which, for the case of nonoverlapping resonances, is given by the Hauser–Feshbach expression. Details of this development are given in Chapters 8, 10, and 11.

As an example we consider here the partial reaction cross-section

$$\sigma_{cc'} = \frac{\pi}{k_c^2} |S_{cc'}|^2 \qquad (1.22a)$$

where k_c is the wave number in channel c and the scattering matrix $S_{cc'}$ for the case of isolated resonances has the form

$$S_{cc'}(E) = -i \sum_\lambda \frac{(\Gamma_{\lambda c}\Gamma_{\lambda c'})^{\frac{1}{2}}}{E - E_\lambda + \frac{i}{2}\Gamma_\lambda}. \qquad (1.22b)$$

By carrying out the averaging process (see Chapter 8), we find the average partial reaction cross-section

$$\langle \sigma_{cc'} \rangle = \frac{\pi}{k_c^2} \frac{T_c T_{c'}}{\sum_{c''} T_{c''}}, \qquad (1.22c)$$

where T_c denotes the quantity $(2\pi\langle\Gamma_{\lambda c}\rangle)/D$, where D is the average spacing. T_c is called the transmission coefficient for channel c. This is the well-known Hauser–Feshbach formula.

As we can easily see compound-nucleus cross-sections provide a vast area for the application of matrix ensembles. Not only the average cross-section but also the fluctuation of the cross-section around its average value can be studied. Only recently an exact expression for the average compound-nucleus cross-section was derived which is valid for all values of the transmission coefficients.

As is clear from the above discussion, in the case of the cross-section and its fluctuation around the mean, we calculate the few low-order moments of the scattering matrix. There are many problems in physics where the low-order moments of some quantity provide much useful information about the system. The best application of low-order moments is to collective nuclear parameters, discussed in Chapters 2–4. The basic idea here is to write the many-body Hamiltonian in terms of angular momentum operators and then calculate the constants appearing in these expressions by calculating the first few moments of the Hamiltonian using a deformed intrinsic many-nucleon wave function. In this way, for example, a new expression for the moment of inertia of the nucleus can be derived. Furthermore, by introducing a correlation coefficient (Kendall 1945) between the many-body Hamiltonian and the square of the total angular momentum operator we can find a relation between this new expression for moment of inertia and the celebrated expression for moment of inertia derived by Skyrme (1957) using a variational procedure.

There are certain other equally interesting applications of matrix ensemble theory which are not described in this book. We only mention some of them.

Matrix ensembles have been used (Rosenzweig *et al.* 1968) to study the violation of time reversal in physical systems. We recall that GOE obeys time reversal invariance while GUE violates time reversal invariance. A new kind of matrix ensemble is introduced, of which one part behaves like GOE and the other part behaves like GUE. The effect of the part of Hamiltonian which violates time reversal on the average and fluctuation properties of the spectra is then studied. In this way upper limits can be put on the violation of time reversal invariance.

Another application of matrix ensemble theory is in the study of thermodynamic properties of the metallic particles (Brody *et al.* 1981).

1.6 Concluding remarks

Before we conclude this introductory chapter we mention briefly the most recent developments and possible new developments in the statistical aspects of many-body physics.

The nuclear many-body grand partition function was studied by Kerman and Troudet (1984) using its functional integral representation. They have studied the expansion of the many-nucleon evolution operator and have derived expressions for nuclear level density.

It is expected that Grassman integration will now make it possible to evaluate exact expressions for the averages of products of four elements of the scattering matrix (Verbaarschot *et al.* 1985). These techniques will also find useful applications in the study of disordered solids.

We should also mention here that the ergodic theorem, which is often assumed to hold good for a number of theoretical derivations in the realm of random matrix ensemble theory, itself needs to be proved rigorously. It is hoped that in future the applicability of the ergodic theorem will be further investigated.

Other possible future developments in the area of many-nucleon problem are: isospin mixing; modification of perturbation theory for the ground and excited states of many-nucleon systems; effects of correlations of E_λ and Γ_λ on the neutron cross-section.

1.7 References

Brody, T. A., Flores, J., French, J. B., Mello, P. A., Pandey, A., and Wong, S. S. M. (1981). *Review of Modern Physics* **53**, 385.

Dyson, F. J. (1962). *Journal of Mathematical Physics* **3,** 1199.
Huizenga, J. R. and Moretto, L. C. (1972). *Annual Reviews of Nuclear Science* **22,** 427.
Kendall, M. G. (1945). *The advanced theory of statistics*. Charles Griffin, London.
Kerman, A. K. and Troudet, T. (1984). *Annals of Physics* (New York) **154,** 456.
Mehta, M. L. (1967). *Random matrices*. Academic Press, New York.
Mehta, M. L. (1971). *Communications in Mathematical Physics* **20,** 245.
Porter, C. E. (1965). *Statistical properties of spectra: fluctuations*. Academic Press, New York.
Rosenzweig, N., Monahan, J. E., and Mehta, M. L. (1968). *Nuclear Physics* **A109,** 437.
Skyrme, T. H. R. (1957). *Proceedings of the Physical Society, London* **A70,** 433.
Verbaarschot, J. J. M., Weidenmüller, H. A., and Zirnbauer, M. R. (1984). *Physical Review Letters* **52,** 1597.
Verbaarschot, J. J. M., Weidenmüller, H. A., and Zirnbauer, M. R. (1985). *Physics Reports* **129,** 367.
Wigner, E. P. (1957). In *Proceedings of the Canadian Mathematical Congress*, p. 174. University of Toronto Press, Toronto.
Wong, S. S. M. (1986). *Nuclear statistical spectroscopy*. Oxford University Press, Oxford.

2
COLLECTIVE ROTATIONAL PARAMETERS OF NUCLEI

2.1 Introduction

Since the early days of nuclear physics it has been known that many heavy nuclei in the region $N \sim 100$, $Z \sim 70$ exhibit rotational bands. A good example (Bohr and Mottelson 1975) is the nucleus ^{168}Er shown in Fig. 2.1. It was later found that the same features are observed in the 2s–1d shell nuclei (Ripka 1968). deVoigt et al. (1983) discussed recent developments in this area.

As is well known, the energy E_J of a given rotational state having angular momentum J is given by

$$E_J = E_0 + \frac{\hbar^2}{2I} J(J+1) \qquad (2.1)$$

where E_0 is called the band head and I the moment of inertia of the nucleus. In this chapter we only discuss even–even nuclei.

The most successful theoretical technique for calculating E_0 and I has been to assume that there is an intrinsic deformed many-body wave function $|\Phi\rangle$ from which the states having a given J can be projected. Many techniques for generating such an intrinsic wave function are given in the literature (Bohr and Mottelson 1975; Ripka 1968). Once such a wave function is obtained, E_0 and I must be calculated using this wave function. We now describe how statistical concepts can be used to calculate E_0 and I given the many-body Hamiltonian H and the many-body intrinsic wave function $|\Phi\rangle$.

2.2 Method of moments

In this section we shall describe the method of moments used to calculate the parameters E_0 and I. We start by writing the operator form of eqn (2.1)

$$H = E_0 + AJ^2 \qquad (2.2)$$

where $A = \hbar^2/2I$, and J^2 is the square of total angular momentum operator.

Writing the first and second moments (Ullah and Sandhya Devi

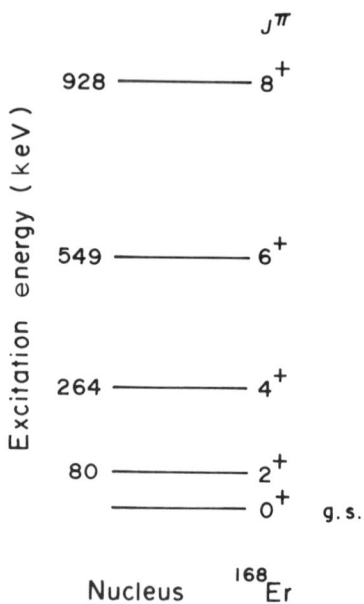

Fig. 2.1 Low-lying rotational levels in the nucleus $^{168}\text{Er}_{100}$. J^π denotes the angular momentum and parity of the level.

1973) of H with respect to $|\Phi\rangle$,

$$\langle H \rangle = E_0 + A\langle J^2 \rangle, \tag{2.3a}$$

$$\langle H^2 \rangle = E_0^2 + 2E_0 A \langle J^2 \rangle + A^2 \langle J^4 \rangle \tag{2.3b}$$

where the bracket sign $\langle \Omega \rangle$ stands for the matrix element $\langle \Phi | \Omega | \Phi \rangle$ of the operator Ω.

Solving eqns (2.3a) and (2.3b) for E_0 and A we get

$$A = \left(\frac{\langle H^2 \rangle - \langle H \rangle^2}{\langle J^4 \rangle - \langle J^2 \rangle^2} \right)^{\frac{1}{2}}, \tag{2.4}$$

$$E_0 = \langle H \rangle - A \langle J^2 \rangle. \tag{2.5}$$

These expressions were also derived by Ng and Trainor (1974) using a different technique.

Since there are two constants E_0 and I in eqn (2.2), we could have used the first moment of the Hamiltonian H and the correlation between H and J^2 which is given by

$$\langle HJ^2 \rangle = E_0 \langle J^2 \rangle + A \langle J^4 \rangle. \tag{2.6}$$

Solving eqns (2.3a) and (2.6) gives

$$A = \frac{\langle HJ^2 \rangle - \langle H \rangle \langle J^2 \rangle}{\langle J^4 \rangle - \langle J^2 \rangle^2}. \tag{2.7}$$

This is the well-known expression for the inverse inertia parameter first obtained by Skyrme (1957) using minimization of $(H - E_0 - AJ^2)^2$.

We call the A obtained by using the method of moments A_M and the one given by Skyrme A_S.

We shall now establish a very simple inequality between A_M and A_S using the correlation coefficient ρ between H and J^2,

$$\rho = \frac{\langle HJ^2 \rangle - \langle H \rangle \langle J^2 \rangle}{[(\langle H^2 \rangle - \langle H \rangle^2)(\langle J^4 \rangle - \langle J^2 \rangle^2)]^{\frac{1}{2}}}. \tag{2.8}$$

Using eqns (2.4) and (2.7)

$$\rho = \frac{A_S}{A_M}. \tag{2.9}$$

Since the correlation coefficient lies between -1 and 1, eqn (2.9) shows that $A_M \geq A_S$.

It is interesting to note from eqn (2.8) that the closeness of ρ to unity implies that

$$(H - \langle H \rangle) \propto (J^2 - \langle J^2 \rangle).$$

2.3 Numerical results

To generate the intrinsic many-nucleon deformed wave function we can either use Nilsson's single-nucleon wave functions (Bohr and Mottelson 1975) or carry out a deformed Hartree–Fock (H–F) calculation. In this section we shall use the H–F wave functions of Ripka (1968) for 2s–1d shell nuclei such as ^{20}Ne and ^{28}Si. In carrying out the H–F calculations it is assumed that these 2s–1d shell nuclei have axial symmetry, and no Coulomb interaction is used. Thus the H–F single-particle orbits are fourfold degenerate. In Table 2.1 we show the oscillator constant and the single-particle energies which were used in these calculations. The two-body interaction was taken to be a Rosenfeld interaction given by

$$V = V_0 \exp\left(-\frac{r}{\mu}\right)^2 \frac{\tau_1 \cdot \tau_2}{3} [0.3 + 0.7\sigma_1 \cdot \sigma_2] \tag{2.10}$$

with $\mu = 1.48$ fm, $V_0 = 70.82$ MeV. The single-particle H–F orbits for the nuclei ^{20}Ne and ^{28}Si are shown in Table 2.2.

The intrinsic deformed H–F wave function $|\Phi\rangle$ for the nucleus ^{20}Ne

14 COLLECTIVE ROTATIONAL PARAMETERS OF NUCLEI

Table 2.1 Values of the oscillator constant $\alpha = (m\omega/\hbar)^{1/2}$ and of the single-particle energies ε_j

Nucleus	Oscillator constant α(fm^{-1})	Single-particle energies (MeV)				
		$\varepsilon_{p_{3/2}}$	$\varepsilon_{p_{1/2}}$	$\varepsilon_{d_{5/2}}$	$\varepsilon_{s_{1/2}}$	$\varepsilon_{d_{3/2}}$
^{20}Ne	0.559	−21.83	−15.67	−4.38	−3.26	0.79
^{28}Si	0.548	−21.83	−15.67	−4.68	−3.26	0.62

Table 2.2 Values of the expansion coefficients c_j^λ for each value of K. For the nucleus ^{28}Si, the oblate solution was used

Nucleus	K	Coefficients c_j^λ				
		$d_{1/2}^{5/2}$	$2s_{1/2}^{1/2}$	$d_{1/2}^{3/2}$	$d_{3/2}^{5/2}$	$d_{3/2}^{3/2}$
^{20}Ne	$\tfrac{1}{2}$	−0.7576	0.5273	0.3847		
^{28}Si	$\tfrac{1}{2}$	−0.5783	−0.7596	0.2977		
	$\tfrac{3}{2}$				0.6935	0.7204
	$\tfrac{3}{2}'$				0.7204	−0.6935

Table 2.3 Values of the energies E_J for the nuclei ^{20}Ne and ^{28}Si. The zero of the energy is taken to be the total H–F energy. The last two columns give the values of E_J using A_S and A_M, respectively

	Nucleus ^{20}Ne			Nucleus ^{28}Si		
	Energies E_J (MeV)			Energies E_J (MeV)		
J	Exact projection	Skyrme	Moments	Exact projection	Skyrme	Moments
0	−3.06	−2.80	−2.81	−2.65	−2.17	−2.39
2	−1.84	−1.77	−1.78	−1.95	−1.57	−1.72
4	0.80	0.63	0.63	−0.35	−0.17	−0.18
6	4.64	4.39	4.42	2.11	2.04	2.24
8	8.02	9.53	9.58	5.30	5.04	5.54

2.4 CORRECTION TO THE $J(J+1)$ SPECTRUM

can be written as

$$|\Phi\rangle = b^+_{\frac{1}{2}p} b^+_{-\frac{1}{2}p} b^+_{\frac{1}{2}n} b^+_{-\frac{1}{2}n} |0\rangle \tag{2.11}$$

where $b^+_{\pm\frac{1}{2}p(n)}$ is the fermion operator that creates a proton(neutron) with spin $\pm\frac{1}{2}$, and $|0\rangle$ represents the ^{16}O core. Similarly we can write down $|\Phi\rangle$ for the nucleus ^{28}Si. Using eqns (2.4), (2.5), and (2.7) we calculate the inverse inertia parameter A and the band head E_0 for these nuclei (Ullah and Sandhya Devi 1973). In Table 2.3 we compare the values of E_J obtained using the statistical values of A and E_0 with those obtained by exact projection from the intrinsic wave function $|\Phi\rangle$. It can be seen from Table 2.3 that the agreement of the approximate values of E_J with their exact values is fairly good. As expected, if we calculate the value of the correlation coefficient ρ given by eqn (2.8), it turns out to be close to unity for both these nuclei, thus indicating that the spectrum is of the $J(J+1)$ type.

2.4 Correction to the $J(J+1)$ spectrum

If the value of the correlation coefficient ρ deviates significantly from unity, we should apply corrections to the expression for E_J. If straightforward expansion in terms of $J(J+1)$ is used, the lowest correction term will be of the form $B[J(J+1)]^2$, where the constant B is much smaller than A. E_J can then be written as

$$E_J = E_0 + AJ(J+1) + B[J(J+1)]^2. \tag{2.12}$$

Holmberg and Lipas (1968) derived a more realistic expression for E_J using a physical picture. They started from the expression

$$E_J = E_0 + \frac{\hbar^2}{2I} J(J+1)$$

and wrote the energy dependence of I as

$$I = I_0 + I_1(E_J - E_0).$$

This gives the expression

$$E_J = E_0 + \frac{\alpha J(J+1)}{1 + (1 + \beta J(J+1))^{\frac{1}{2}}} \tag{2.13}$$

where α and β are constants.

From eqns (2.12) and (2.13) we see that we now have to determine one more constant, B or β. This can easily be done by the method of moments by calculating the third-order moment $\langle H^3 \rangle$ of the Hamiltonian. We note that, since eqn (2.12) involves only J^2 and J^4, its moments will be much simpler to calculate than the square root function $(1+J^2)^{\frac{1}{2}}$. Equation (2.13), however, gives a much better fit to the observed spectra than eqn (2.12). We shall further discuss the calculation of such quantities in Chapter 7. A similar treatment can be used to determine the parameters of other models which have been put forward from time to time (de Voigt et al. 1983) to correct the simple expression for E_J given in eqn (2.1).

2.5 Semi-classical formulation

Bhaduri and Das Gupta (1973) presented a different approach based on partition functions for studying the rotational parameters of the nucleus. In this section we give a simplified version of this approach. The starting point is to write the partition function $z(\beta)$ for the rotational Hamiltonian as

$$z(\beta) = \sum_{J=0, 2, \ldots} (2J+1)\exp\left[-\frac{\hbar^2}{2I}\beta J(J+1)\right] \quad (2.14)$$

where we consider an even–even nucleus. $\beta = 1/(kT_B)$, where T_B is the Boltzmann temperature. The zero of the energy is taken to be the ground state $J = 0$.

Now the average value of J^2 can be written in two ways. First,

$$\langle J^2 \rangle = \sum a_J^2 J(J+1) \quad (2.15)$$

where the a_Js, assumed to be real, are expansion coefficients which express $|\Phi\rangle$ in terms of $|\Psi_J\rangle$. In the second formulation,

$$\langle J^2 \rangle = \frac{\sum J(J+1)(2J+1)\exp\left(-\frac{\hbar^2}{2I}\beta J(J+1)\right)}{\sum (2J+1)\exp\left(-\frac{\hbar^2}{2I}\beta J(J+1)\right)}. \quad (2.16)$$

Equations (2.15) and (2.16) suggest that a_J^2 can be expressed as

$$a_J^2 = K(2J+1)\exp[-\lambda J(J+1)] \quad (2.17)$$

where K and λ are constants.

We determine these constants using the requirements

$$\sum a_J^2 = 1, \quad (2.18a)$$

$$\sum a_J^2 J(J+1) = \langle J^2 \rangle. \quad (2.18b)$$

Putting eqn (2.17) into (2.18a, b) and replacing the sums by integrals, we can easily show that

$$a_J^2 = \frac{2}{\langle J^2 \rangle}(2J+1)\exp\left(-\frac{J(J+1)}{\langle J^2 \rangle}\right). \tag{2.19}$$

It is a simple consequence of eqn (2.19) that

$$\langle J^4 \rangle = 2\langle J^2 \rangle^2. \tag{2.20}$$

To calculate the inertia parameter I, we use the statistical relation

$$\langle H^2 \rangle - \langle H \rangle^2 = -\frac{\partial}{\partial \beta}\langle H \rangle. \tag{2.21}$$

Since

$$\langle H \rangle = A\langle J^2 \rangle, \tag{2.22a}$$

and, from eqn (2.16),

$$\langle J^2 \rangle = \frac{1}{A\beta}, \tag{2.22b}$$

(2.22b) combined with (2.21) give the following expression for the inverse of the inertia parameter A,

$$A = \frac{(\langle H^2 \rangle - \langle H \rangle^2)^{\frac{1}{2}}}{\langle J^2 \rangle}. \tag{2.23}$$

We note that this is the same expression as the one given by the method of moments, provided that $\langle J^4 \rangle$ is further approximated by $2\langle J^2 \rangle^2$.

2.6 References

Bhaduri, R. K. and Das Gupta, S. (1973). *Nuclear Physics* **A212,** 18.
Bohr, A. and Mottelson, B. R. (1975). *Nuclear structure,* Vol. II. Benjamin, (Reading) Massachusetts.
deVoigt, M. J. A., Dudek, J., and Szymanski, Z. (1983). *Review of Modern Physics* **55,** 949.
Holmberg, P. and Lipas, P. O. (1968). *Nuclear Physics* **A117,** 552.
Ng, W. and Trainor, L. E. H. (1974). *Canadian Journal of Physics* **52,** 541.
Ripka, G. (1968). In *Advances in nuclear physics* (ed. M. Baranger and E. Vogt), Vol. I, p. 183. Plenum Press, New York.
Skyrme, T. H. R. (1957). *Proceedings of the Physical Society, London* **A70,** 433.
Ullah, N. and Sandhya Devi, K. R. (1973). *Physical Review* **C8,** 1167.

3
VIBRATIONAL NUCLEI

3.1 Introduction

The heavy nuclei which are almost spherical in shape exhibit shape oscillations just like the small oscillations of a liquid drop. The collective vibrational Hamiltonian for small oscillations can be written (Rowe 1970) as

$$H = \sum_{\lambda\mu} \hbar\omega_\lambda (O^+_{\lambda\mu} O_{\lambda\mu} + \tfrac{1}{2}) \tag{3.1}$$

where the operator $O^+_{\lambda\mu}$ creates a phonon of angular momentum λ and its projection quantum number μ. Thus, the vibrational states are labelled by the number of phonons and the possible values of angular momenta J to which these phonons can couple.

For illustrative purposes we consider quadrupole phonons: then $\lambda = 2$ and the first excited state will be a one-phonon state having $J = 2$. The second excited state will have two phonons and the possible values of J are 0, 2, and 4. A good example of such a vibrational spectrum is the nucleus ^{114}Cd shown in Fig. 3.1.

The excited vibrational states are usually described by the random phase approximation (RPA) theory. The complete description of RPA can be found in any standard textbook on nuclear theory. Here we shall give a brief outline of the theory which is needed later in this chapter.

The basic idea is to expand the phonon creation operator O^+_λ in terms of particle–hole operators (Rowe 1970)

$$O^+_\lambda = \sum_{mi} [Y_{mi}(\lambda) a^+_m a_i - Z_{mi}(\lambda) a^+_i a_m] \tag{3.2}$$

where a^+ and a are particle creation and annihilation operators and m is used to denote an unoccupied single-particle H–F state and i an occupied state. It is assumed that the ground state can be approximated by a H–F particle–hole vacuum (Rowe 1970) and Y and Z denote the amplitudes for the creation and destruction of a particle–hole pair, respectively. One can then derive an equation describing the process using either a variational or linearization approximation for these amplitudes. This is the RPA equation and is given (Rowe 1970) by

$$\begin{pmatrix} A & B \\ B^* & A^* \end{pmatrix} \begin{pmatrix} \mathbf{Y}(\lambda) \\ \mathbf{Z}(\lambda) \end{pmatrix} = \hbar\omega_\lambda \begin{pmatrix} \mathbf{Y}(\lambda) \\ -\mathbf{Z}(\lambda) \end{pmatrix} \tag{3.3}$$

3.2 DESCRIPTION IN TERMS OF AN INTRINSIC STATE

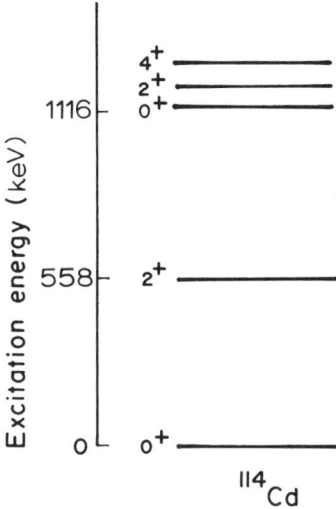

Fig. 3.1 Vibrational spectrum observed in the nucleus ^{114}Cd.

where $\mathbf{Y}(\lambda)$ and $\mathbf{Z}(\lambda)$ are vectors, the components of which are $Y_{mi}(\lambda)$ and $Z_{mi}(\lambda)$, the submatrices A and B are defined by

$$A_{minj} \equiv \langle |[a_i^+ a_m, H, a_n^+ a_j]| \rangle, \tag{3.4a}$$

$$B_{minj} \equiv -\langle |[a_i^+ a_m, H, a_j^+ a_n]| \rangle, \tag{3.4b}$$

and ω_λ are the RPA eigenvalues.

3.2 Description in terms of an intrinsic state

We would now like to see whether there is some way in which collective vibrational states can also be described using the concept of an intrinsic deformed state. This can be done if we first take only the state of maximum angular momentum from each excited state and consider the states $J = 0, 2, 4, \ldots$ having zero, one, two ... phonons as the members of a given band of states. We assume that there is an underlying intrinsic state from which we can project these states having $J = 0, 2, 4, \ldots$.

In a simple model Haapakoski et al. (1970) showed that such a state is a coherent phonon state. A more realistic way of constructing an intrinsic deformed state was given by Gupta and Ullah (1971). According to this later prescription one first constructs a deformed H–F state and then carries out an RPA calculation of the type described in § 3.1 but in which the operator O_λ^+ is expanded in terms of intrinsic particle–hole states. The states $J = 0, 2, 4, \ldots$ are then projected from a many-nucleon intrinsic state obtained using a method due to Sanderson (1965) which makes use of a quasi-boson approximation. More details are given in a later section in which we describe how to calculate ω, the RPA eigenvalues, using the method of moments. We next describe the essential basic method formulation for calculating ω.

3.3 Derivation of an expression for ω

Taking the analogy from the rotational case, we write the vibrational energies

$$E_J = E_0 + \omega J \tag{3.5}$$

where E_0 is the energy corresponding to $J = 0$ and 2ω gives the spacing.

In the case of rotational spectra, it was straightforward to write the operator form of the expression for the energies E_J. For vibrational states we first need to construct an operator (Ullah 1982) which has eigenvalues J. Let Ω be such an operator; then

$$\Omega |\Psi_J\rangle = J |\Psi_J\rangle. \tag{3.6}$$

Since we know that

$$J^2 |\Psi_J\rangle = J(J+1) |\Psi_J\rangle, \tag{3.7}$$

we operate on both sides of eqn (3.6) with Ω and write

$$\Omega^2 |\Psi_J\rangle = J^2 |\Psi_J\rangle. \tag{3.8}$$

Combining eqns (3.6)–(3.8) we get the operator relation

$$[\Omega^2 + \Omega - J^2] |\Psi_J\rangle = 0. \tag{3.9}$$

This shows that the operator Ω is given by

$$\Omega = -\tfrac{1}{2} + \tfrac{1}{2}(1 + 4J^2)^{\frac{1}{2}}. \tag{3.10}$$

The sign of the square root is chosen to obtain the correct value of Ω when it operates on the state $|\Psi_{J=0}\rangle$.

It is now a simple matter to write eqn (3.5) in the operator form

$$H = E_0 + \omega \Omega. \tag{3.11}$$

Let $|\chi\rangle$ be the underlying intrinsic state; then, as in the rotational case, we can take the moments of H and derive the expressions

$$\omega = \left(\frac{\langle H^2 \rangle - \langle H \rangle^2}{\langle \Omega^2 \rangle - \langle \Omega \rangle^2} \right)^{\frac{1}{2}}, \tag{3.12a}$$

$$E_0 = \langle H \rangle - \omega \langle \Omega \rangle \tag{3.12b}$$

3.4 Peierls–Yoccoz type of formulation for the vibrational parameters

where $\langle \ \rangle$ denotes the matrix element of the operator with respect to $|\chi\rangle$.

The Hill–Wheeler integral (Hill and Wheeler 1953) was used by Peierls and Yoccoz (1957) to study the collective states of nuclei. We now show how this formulation can be used along with a statistical approximation to derive expressions for E_0 and ω.

Let $|\chi\rangle$ be the intrinsic deformed state from which we project states $J = 0, 2, 4, \ldots$; then the energies E_J can be written (Peierls and Yoccoz 1957) as

$$E_J = \frac{\int_0^\pi d\beta \sin\beta \, \langle \chi | H \exp(-i\beta J_y) | \chi \rangle \, P_J(\cos\beta)}{\int_0^\pi d\beta \sin\beta \, \langle \chi | \exp(-i\beta J_y) | \chi \rangle \, P_J(\cos\beta)} \tag{3.13}$$

where H is the Hamiltonian, J_y the y-component of the angular momentum operator, and P_J a Legendre polynomial. Since we are considering even values of J, the integral from $\pi/2$ to π is the same as from 0 to $\pi/2$. Writing

$$H = \langle \chi | H | \chi \rangle + (H - \langle \chi | H | \chi \rangle) \tag{3.14}$$

and

$$\cos\beta = x,$$

eqn (3.13) gives

$$E_J = \langle H \rangle + \frac{\int_0^1 dx P_J(x) h(x)}{\int_0^1 dx P_J(x) n(x)} \tag{3.15}$$

where

$$h(x) = \sum a_J^2 (E_J - \langle H \rangle) P_J(x) \tag{3.16a}$$

and

$$n(x) = \sum a_J^2 P_J(x) \tag{3.16b}$$

where a_J are the expansion coefficients.

From the formulation of § 3.3 we see that Ω plays the same role in the vibrational case as J^2 played in the rotational case. We now expand P_J around the average value of J

$$P_J(x) = P_{\langle\Omega\rangle}(x) + (J - \langle\Omega\rangle)\left[\frac{\partial}{\partial v}P_v(x)\right] + \ldots, \qquad v = \langle\Omega\rangle \qquad (3.17)$$

where

$$\langle\Omega\rangle = \langle\chi|\Omega|\chi\rangle = \sum a_J^2 J. \qquad (3.18)$$

In writing eqn (3.17) we used a Taylor series expansion around $\langle\Omega\rangle$ and introduced the generalized Legendre function P_v (Gradshteyn and Ryzhik 1965).

Using expansion (3.17) in eqn (3.16) we can rewrite the energies E_J given by eqn (3.15) as

$$E_J = \langle H\rangle + V\frac{\partial}{\partial v}\ln\left[\int_0^1 dx P_v(x)P_J(x)\right], \qquad v = \langle\Omega\rangle \qquad (3.19)$$

where

$$V = \langle\chi|(H - \langle H\rangle)(\Omega - \langle\Omega\rangle)|\chi\rangle. \qquad (3.20)$$

Let I_v denote the integral

$$I_v = \int_0^1 dx P_v(x) P_J(x). \qquad (3.21)$$

Then, using expansion (3.17) for P_J, we get

$$I_v = \int_0^1 dx P_v(x) P_{\langle\Omega\rangle}(x)$$
$$+ (J - \langle\Omega\rangle)\int_0^1 dx P_v(x)\left[\frac{\partial}{\partial \mu}P_\mu(x)\right], \qquad \mu = \langle\Omega\rangle. \qquad (3.22)$$

Using eqns (3.19), (3.21), and (3.22), we can write E_J, after some simplification, as

$$E_J = \left(\langle H\rangle - V\langle\Omega\rangle\frac{\partial^2}{\partial v\,\partial\mu}\ln B_{\mu v} + V\frac{\partial}{\partial v}\ln B_{\mu v}\right)$$
$$+ \left(V\frac{\partial^2}{\partial v\,\partial\mu}B_{\mu v}\right)J \qquad (3.23)$$

where $B_{\mu v}$ denotes the integral

$$B_{\mu v} = \int_0^1 dx P_\mu(x) P_v(x). \qquad (3.24)$$

In eqn (3.23) after the indicated differentiation we put $\mu = v = \langle\Omega\rangle$.

Expression (3.23) is the desired expression for E_J; it is of the form

$$E_J = E_0 + \omega J. \tag{3.25}$$

Thus ω is given by

$$\omega = V \left(\frac{\partial^2}{\partial v \, \partial \mu} \ln B_{\mu v} \right)_{\mu = v = \langle \Omega \rangle}. \tag{3.26}$$

The exact value of $B_{\mu v}$ is known (Gradshteyn and Ryzhik 1965) but is quite complicated. For our purpose, since $\langle \Omega \rangle$ is fairly large and we wish $\mu = v = \langle \Omega \rangle$, we can approximately write

$$B_{\mu v} = (\mu + v + 1)^{-1}. \tag{3.27}$$

This gives

$$\omega = \frac{\langle (H - \langle H \rangle)(\Omega - \langle \Omega \rangle) \rangle}{4 \langle \Omega \rangle^2}. \tag{3.28}$$

It is interesting to note here that we can also calculate ω by using 'Skyrme's type of expression for ω' and calculate $\langle \Omega^2 \rangle - \langle \Omega \rangle^2$ using the semi-classical approximation described in Chapter 2. This gives

$$\omega_s = \frac{\langle (H - \langle H \rangle)(\Omega - \langle \Omega \rangle) \rangle}{0.5 \langle \Omega \rangle^2}. \tag{3.29}$$

Comparing eqns (3.28) and (3.29), we find that the ω given by Skyrme's procedure is much larger than that given by Peierls–Yoccoz formulation.

3.5 Model calculation

In this section we apply the formulation developed in § 3.3 to a simple model which describes the $J = 0, 2, 4$ states of the nucleus ^{18}O. Taking ^{16}O to be the core, we use the single-particle energies (Gupta and Ullah 1971) ε_j,

$$\varepsilon_{d_{\frac{5}{2}}} = -4.50 \text{ MeV} \quad \text{and} \quad \varepsilon_{2s_{\frac{1}{2}}} = -3.28 \text{ MeV}.$$

The two-body interaction is taken to be a 40-MeV Rosenfeld interaction.

As a first step in generating the intrinsic deformed state which has $J = 0, 2, 4$ states, we carry out the H–F calculation which gives the lowest deformed orbit.

$$b^+_{k=\frac{1}{2}} = 0.980 a^+_{1d\frac{5}{2}\frac{1}{2}} - 0.199 a^+_{2s\frac{1}{2}\frac{1}{2}},$$

which has fourfold degeneracy.

Second, the intrinsic RPA calculation is performed giving the RPA amplitudes and frequencies

$$\omega_1 = 2.0630, \quad \begin{pmatrix} y_1^1 \\ y_2^1 \end{pmatrix} = \begin{pmatrix} 0.7096 \\ -0.7096 \end{pmatrix}, \quad \begin{pmatrix} z_1^1 \\ z_2^1 \end{pmatrix} = \begin{pmatrix} -0.0591 \\ 0.0591 \end{pmatrix};$$

$$\omega_2 = 1.8853, \quad \begin{pmatrix} y_1^2 \\ y_2^2 \end{pmatrix} = \begin{pmatrix} 0.7101 \\ 0.7101 \end{pmatrix}, \quad \begin{pmatrix} z_1^2 \\ z_2^2 \end{pmatrix} = \begin{pmatrix} -0.0646 \\ 0.0646 \end{pmatrix}.$$

It is now easy to construct the intrinsic Sanderson state $|\chi\rangle$ given by

$$|\chi\rangle = N_0 \exp\left[-\tfrac{1}{2} \sum C_{\alpha\beta} \eta_\alpha^+ \eta_\beta^+\right] |\Phi_0\rangle \qquad (3.30)$$

where N_0 is the normalization constant and the correlation matrix C is given by $z = Cy$. In (3.30) η_α^+ is the operator $b_m^+ b_i$ ($\alpha \equiv mi$) and creates a deformed particle–hole state.

We now calculate ω and E_0 using eqns (3.12a) and (3.12b). Their values turn out to be $\omega = 2.162$ MeV and $E_0 = -11.860$ MeV which compare fairly well with the values of $\omega = 1.523$ MeV and $E_0 = -11.118$ MeV extracted from the exactly projected $J = 0$ and $J = 2$ states using the intrinsic state given by eqn (3.30).

3.6 References

Gradshteyn, I. S. and Ryzhik, I. M. (1965). In *Tables of integrals, series and products* (ed. A. Jeffrey). 794. Academic Press, New York and London.

Gupta, K. K. and Ullah, N. (1971). *Physics Letters* **B36,** 196.

Haapakoski, P., Hankaranta, T., and Lipas, P. O. (1970). *Physics Letters* **31B,** 493.

Hill, D. L. and Wheeler, J. A. (1953). *Physical Review* **89,** 1102.

Peierls, R. E. and Yoccoz, J. (1957). *Proceedings of the Physical Society, London* **A70,** 381.

Rowe, D. J. (1970). *Nuclear collective motion*, Chapter 2. Methuen, London.

Sanderson, E. (1965). *Physics Letters* **19,** 141.

Ullah, N. (1982). *Pramana* **18,** 211.

4

TRANSITION REGION IN HEAVY NUCLEI

4.1 Introduction

Reactions in which a heavy ion bombards a heavy nucleus and a number of neutrons are emitted (Ward *et al.* 1968) are of great current interest. In these reactions we study the energy levels of neighbouring neutron-deficient isotopes. A simple example is shown in Fig. 4.1 where the energy levels of Ce isotopes are given. We see from Fig. 4.1 that the nucleus ^{136}Ce has almost constant spacing of energy levels while the nulceus ^{128}Ce has energy level spacing proportional to $J(J+1)$ where J is the total angular momentum of the level. Between these two nuclei, nuclei such as ^{132}Ce have neither constant spacing of energy levels nor is the spacing proportional to $J(J+1)$. From Chapter 3 we know that spherical nuclei can exhibit a vibrational spectrum, whereas, if the nucleus is fully deformed, we get a rotational spectrum as discussed in Chapter 2. Thus the obvious conclusion is that nuclei like ^{132}Ce are neither spherical nor fully deformed and that, therefore, their spectra can be described neither in terms of pure vibrations nor of pure rotations. In this region the vibrational and rotational modes start interacting with each other.

In a phenomenological study Ejiri and his collaborators (1968) found that, if $(E_J - E_0)/J$ is plotted versus J, the spectra of heavy nuclei like those shown in Fig. 21.1 fall almost on a straight line. Thus the energies E_J can be written

$$E_J = E_0 + AJ + BJ(J+1) \tag{4.1}$$

where E_0, A, and B are constants.

From the theoretical point of view it is interesting to see whether we can develop a formalism which describes the behaviour of E_J as it goes from the vibrational to the rotational limit. In the next section we describe a simple model put forward by Haapakoski *et al.* (1970). In § 4.3 we give a different formulation based on the correlation coefficient between the many-body Hamiltonian and the projection operator. It is shown in § 4.4 that the same expression can be derived using Bohr–Mottelson's theory. In § 4.5 we shall fit the spectra of the ^{152}Gd nucleus using the expressions given in §§ 4.2 and 4.3.

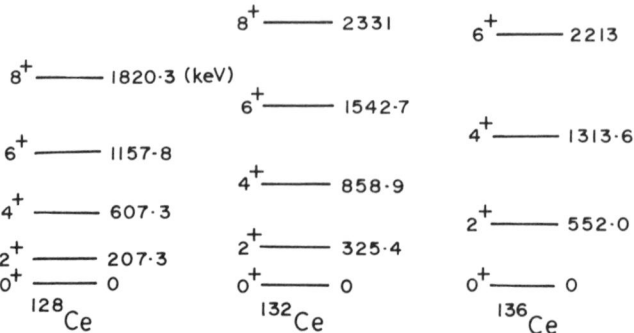

Fig. 4.1 Spectra of the isotopes of Ce nucleus.

4.2 Coherent phonon model

As in the earlier chapters we are interested in constructing an underlying intrinsic state which can be used to project good angular momentum states as we go from the vibrational limit to the rotational limit. Haapakoski et al. (1970) showed that the proper underlying intrinsic state is given by

$$|\tilde{0}\rangle = N \exp(db_{20}^+) |0\rangle \qquad (4.2)$$

where b_{20}^+ denotes a spherical phonon creation operator with $l = 2$, $m = 0$, $|0\rangle$ denotes the spherical phonon vacuum state, N is the normalization constant and d the deformation parameter. The intrinsic state given by eqn (4.1) is also the coherent phonon state as can be easily checked by operating on it with the annihilation operator b_{20}.

Approximating the Hamiltonian by

$$H = C_1 \sum_m b_{2m}^+ b_{2m} \qquad (4.3a)$$

where C_1 is a constant, and using the usual technique of angular momentum projection, it can be shown (Haapakoski et al. 1970) that the energies E_J are given by

$$E_J = C_1 d^2 \frac{\int_0^1 dx P_J(X) P_2(X) \exp[d^2 P_2(X)]}{\int_0^1 dx P_J(X) \exp[d^2 P_2(X)]} . \qquad (4.3b)$$

As the deformation parameter d varies from 0 to ∞, the spectrum E_J given by eqn (4.3b) reproduces the features shown in Fig. 4.1.

In the next section we derive an energy expression which makes use

4.3 Energy expression in the rotation–vibration region

Let us assume that there is an intrinsic state $|\chi\rangle$ from which we can project the states of good angular momentum $|\Psi_J\rangle$. In the simplest approximation $|\chi\rangle$ could be the coherent phonon state given by eqn (4.2). A more realistic $|\chi\rangle$ can be obtained by first carrying out a deformed H–F calculation and then carrying out an RPA calculation using the deformed H–F state as the approximate RPA vacuum. Let H be the total Hamiltonian of the system and P_J the projection operator which projects $|\psi_J\rangle$ from $|\chi\rangle$. Then the correlation coefficient between H and P_J can be written (Ullah 1981) as

$$\rho_J = \frac{\langle \chi| HP_J |\chi\rangle - \langle \chi| H |\chi\rangle \langle \chi| P_J |\chi\rangle}{[(\langle \chi| H^2 |\chi\rangle - \langle \chi| H |\chi\rangle^2)(\langle \chi| P_J^2 |\chi\rangle - \langle \chi| P_J |\chi\rangle^2)]^{\frac{1}{2}}}. \tag{4.4}$$

The projection operator P_J can formally be written

$$P_J = |\Psi_J\rangle \langle \Psi_J|. \tag{4.5a}$$

It satisfies the usual relation

$$P_J^2 = P_J. \tag{4.5b}$$

The energy E_J of the state $|\Psi_J\rangle$ is given by

$$E_J = \frac{\langle \chi| HP_J |\chi\rangle}{\langle \chi| P_J |\chi\rangle}. \tag{4.6}$$

Combining eqns (4.4)–(4.6),

$$E_J = \langle H\rangle + (\langle H^2\rangle - \langle H\rangle^2)^{\frac{1}{2}} \rho_J (\langle P_J\rangle^{-1} - 1)^{\frac{1}{2}}, \tag{4.7}$$

where the bracket sign $\langle \ \rangle$ denotes the matrix element with respect to $|\chi\rangle$. Denoting $\rho_J(\langle P_J\rangle^{-1} - 1)^{\frac{1}{2}}$ by X_J,

$$E_J = \langle H\rangle + (\langle H^2\rangle - \langle H\rangle^2)^{\frac{1}{2}} X_J. \tag{4.8}$$

We now use the following equations to parametrize E_J,

$$\sum a_J^2 E_J = \langle H\rangle, \tag{4.9a}$$

$$\sum a_J^2 E_J^2 = \langle H^2\rangle, \tag{4.9b}$$

$$\sum a_J^2 J(J+1) E_J = \langle HJ^2\rangle. \tag{4.9c}$$

Putting eqn (4.8) into (4.9),

$$\sum a_J^2 X_J = 0, \qquad (4.10a)$$

$$\sum a_J^2 X_J^2 = [\langle H^2 \rangle - \langle H \rangle^2]^{-\frac{1}{2}}, \qquad (4.10b)$$

$$\sum a_J^2 J(J+1) X_J = \frac{\langle HJ^2 \rangle - \langle H \rangle \langle J^2 \rangle}{(\langle H^2 \rangle - \langle H \rangle^2)^{\frac{1}{2}}}. \qquad (4.10c)$$

Multiplying eqns (4.9a, b, c) by constants k_1, k_2 and k_3, respectively and adding them, we can write

$$\sum a_J^2 [k_1 X_J + k_2 X_J^2 + k_3 J(J+1) X_J]$$

$$= [\langle H^2 \rangle - \langle H \rangle^2]^{-\frac{1}{2}} k_2 + \frac{\langle HJ^2 \rangle - \langle H \rangle \langle J^2 \rangle}{(\langle H^2 \rangle - \langle H \rangle^2)^{\frac{1}{2}}} k_3. \qquad (4.11)$$

The constants k_1, k_2 and k_3 can later be determined in terms of $\langle H \rangle$, $\langle H^2 \rangle$, and $\langle HJ^2 \rangle$ by using eqns (4.9).

Since $\sum a_J^2 = 1$ and assuming that eqn (4.11) holds for each J separately, we can write X_J as the solution of the quadratic equation. Since our purpose was to parametrize E_J, we can redefine the constants and from eqns (4.11) and (4.8) we can write E_J approximately as

$$E_J = E_0 + AJ(J+1) + (B^2(J+\tfrac{1}{2})^2 + C)^{\frac{1}{2}} \qquad (4.12)$$

where A, B, and C are constants.

Realizing that the parameter B plays the same role in the vibrational case as A plays in the rotational case, we can write

$$E_J = E_0 + AJ(J+1) + \left(B^2(J+\tfrac{1}{2})^2 + \frac{A}{4}(A+2B)\right)^{\frac{1}{2}}. \qquad (4.13)$$

If desired, E_0, A, and B can now be determined in terms of $\langle H \rangle$, $\langle H^2 \rangle$, and $\langle HJ^2 \rangle$ as in the last two chapters. If b becomes much larger than A, we can expand the square root and derive Ejiri's expression, eqn (4.1).

4.4 Energy expression based on Bohr–Mottelson's collective Hamiltonian

The collective properties of heavy nuclei are well described by Bohr–Mottelson's (1953) collective Hamiltonian. For the case of rotations and

β-vibrations, this Hamiltonian can be written (Davidson 1968) as

$$H = \frac{J^2}{2I} + \tfrac{1}{2}C(\beta - \beta_0)^2, \tag{4.14}$$

where I is the inertia parameter of the nucleus, C the spring constant, and β_0 the equilibrium value of β. The energies denoted by E_{J,n_β} are then given (Ullah 1983) by

$$E_{J,n_\beta} = \frac{J(J+1)}{2I} + \left(\frac{C}{B}\right)^{\frac{1}{2}}(n_\beta + \tfrac{1}{2}) \tag{4.15}$$

where n_β denotes the number of phonons and B is the mass parameter.

If we now introduce an interaction term $\lambda J^2(\beta - \beta_0)^2$, λ being a constant, which takes into account the interaction between rotational and vibrational modes, the Hamiltonian H given by eqn (4.14) becomes

$$H = \frac{J^2}{2I} + \tfrac{1}{2}C(\beta - \beta_0)^2 + \lambda J^2(\beta - \beta_0)^2. \tag{4.16}$$

Considering the ground-state band $n_\beta = 0$, we see from eqn (4.16) that the energies E_J in the ground-state band are

$$E_J = \frac{J(J+1)}{2I} + \left(\frac{\lambda}{2B}(J+\tfrac{1}{2})^2 + \frac{1}{4B}\left(C - \frac{\lambda}{2}\right)\right)^{\frac{1}{2}}. \tag{4.17}$$

This is the same kind of expression which we derived in § 4.3 using the correlation coefficient between the total Hamiltonian and the angular momentum projection operator.

Since the parameters of eqn (4.13) are known in terms of matrix elements of $\langle H \rangle$, $\langle H^2 \rangle$, and $\langle HJ^2 \rangle$ we can, therefore, express the parameters I, B, and λ of the Bohr–Mottelson collective Hamiltonian in terms of the matrix elements of the many-body Hamiltonian with respect to the intrinsic wave function.

4.5 Fitting of the experimental data

Since there have been no microscopic calculations to generate the appropriate intrinsic wave function in the region where rotational and vibrational modes interact we use eqn (4.13) to fit the experimental spectrum of a nucleus which is neither spherical nor fully deformed. We now show how the fit compares with eqn (4.3b) which was obtained using a simple model. Rather than doing a χ^2 minimization to obtain the parameters A and B, we determine them from the first three levels of the nucleus ^{152}Gd as was done by Haapakoski et al. (1970). The remaining levels are then fitted using these parameters. The results are shown in

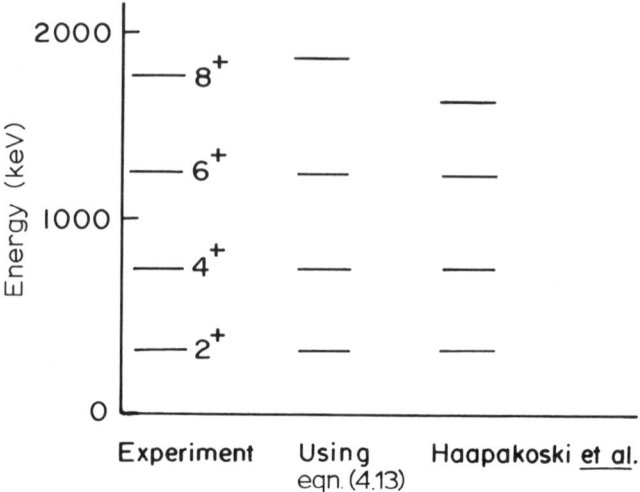

Fig. 4.2 Experimental spectra of ^{152}Gd compared with the spectra obtained using eqn (4.13) and with the model put forward by Haapakoski *et al.* (1970).

Fig. 4.2. We see that eqn (4.13) gives a much improved fit to the observed experimental spectrum compared with the one obtained using a simple model.

4.6 Concluding remarks

In Chapters 2, 3, and 4 we described some of the common collective models of nuclei which one uses in fitting the observed spectra. The expressions are quite simple for either pure rotations or pure vibrations but slightly more complicated when there is interaction between rotational and vibrational modes. The collective parameters in all cases can be obtained by the method of moments using an underlying deformed intrinsic wave function. The coherent phonon state of Haapakoski *et al.* (1970) given by eqn (4.2) can further be generalized to include phonon operators with higher l values. These developments go far beyond the scope of the present book and are therefore not described here.

4.7 References

Bohr, A. and Mottelson, B. R. (1953). *Kongelige Danske Videnskabernes Selskab, Matematisk-Fysiske Meddelelser* **27**(16).
Davidson, J. P. (1968). *Collective models of the nucleus*, Chapter 3. Academic Press, New York.

4.7 REFERENCES

Ejiri, H., Ishihara, M., Sakai, M., Katori, K., and Inamura, T. (1968). *Journal of the Physical Society of Japan* **24,** 1189.
Haapakoski, P., Honkaranta, T., and Lipas, P. O. (1970). *Physics Letters* **31B,** 493.
Ullah, N. (1981). *Journal of Physics* **G7,** L149.
Ullah, N. (1983). *Journal of the Physical Society of Japan* **52,** 357.
Ward, D., Diamond, R. M., and Stephens, F. S. (1968). *Nuclear Physics* **A117,** 309.

5
SYMMETRY MIXING

5.1 Introduction

The wave functions of a many-body system possess certain symmetries; for example, in the light atoms the wave functions can be classified by total orbital angular momentum L and total spin quantum number S. As the spin–orbit interaction increases L and S separately do not remain good quantum numbers and we talk about the mixing of two symmetries, namely the orbital and the spin symmetry. Another familiar example is the ground state of the deuteron which has $J^\pi = 1^+$, J being the angular momentum and π the parity. If there were no tensor force, the deuteron's ground state would be a $^3S_1$ state with $L = 0$ and $S = 1$. The tensor force mixes L and S and the ground state becomes a mixture of $^3S_1$ and $^3D_1$ states. Similarly, because of the Coulomb interaction in the many-nucleon system the isotopic spin quantum number T does not remain a good quantum number in going from light nuclei to medium nuclei. The shell-model picture of the nucleus provides a further example. Here, because of nuclear interaction, the pure shell-model configurations are mixed together.

Information about mixing of symmetries is useful in constructing the wave functions of a many-body system. Even if the symmetries are not pure, we can greatly reduce the size of the interaction matrix in calculating the relevant eigenstates and eigenvalues of the system. In this chapter we would like to describe a method which uses certain average values of the Hamiltonian matrix elements to get information about mixing of symmetries. In § 5.2 we shall give a derivation of the mixing parameter and show its applications in § 5.3.

5.2 Derivation of the expression for the mixing parameter

Let us consider a light nucleus such as ^{14}N which has seven protons and seven neutrons. In the zero-order shell model picture (Preston 1962; Preston and Bhaduri 1975) its ground state will be a pure configuration $(p_\frac{1}{2})^2$ with the two $p_\frac{1}{2}$ nucleons coupled to $J^\pi = 1^+$, $T = 0$. Because of nuclear interaction this configuration will mix, e.g. with the configuration $(p_\frac{3}{2})^{-1} (p_\frac{1}{2})^{-1}$ in which there is one hole in $p_\frac{3}{2}$ and one hole in $p_\frac{1}{2}$. In such cases we are interested in knowing the amount of mixing of this

5.2 EXPRESSION FOR THE MIXING PARAMETER

configuration with the dominant configuration $(p_{\frac{1}{2}})^2$. Wave functions having a given value of J^π and T and belonging to a particular configuration are denoted ϕ_α. In general there are many ϕ_αs having the same value of J^π and T and belonging to the same configuration. This will be denoted by $\bar{\alpha}$. Thus $\bar{\alpha}$ will specify a particular symmetry. We assume that the Hamiltonian has terms which bring about the mixing of symmetries $\bar{\alpha}$, $\bar{\beta}$, etc. The problem of mixing of symmetries becomes easier to treat if, following French (1967) and Chang (1970), we introduce certain quantities called centroids and partial widths. The centroid $\varepsilon(\bar{\alpha})$ for a symmetry $\bar{\alpha}$ is defined as

$$\varepsilon(\bar{\alpha}) = d^{-1}(\bar{\alpha}) \sum_{\alpha \in \bar{\alpha}} \langle \phi_\alpha | H | \phi_\alpha \rangle \tag{5.1}$$

where $d(\bar{\alpha})$ is the dimension of the symmetry $\bar{\alpha}$.

The square of the partial width $\sigma(\bar{\alpha}, \bar{\beta})$ is defined by

$$\sigma^2(\bar{\alpha}, \bar{\beta}) = d^{-1}(\bar{\alpha}) \sum_{\substack{\alpha \in \bar{\alpha} \\ \beta \in \bar{\beta}}} \langle \phi_\alpha | H | \phi_\beta \rangle^2. \tag{5.2}$$

In writing definition (5.2) we assume that the matrix elements of the Hamiltonian are real. Special techniques have been developed to calculate the average matrix elements (Wong 1986) which enter into eqns (5.1) and (5.2).

Now consider the mixing of two symmetries, $\bar{\alpha}$ and $\bar{\beta}$. Let ψ_j be the exact eigenfunction of the Hamiltonian H. Then it can be written as (Ullah 1971)

$$\psi_j = \lambda \left[\sum_{\alpha \in \bar{\alpha}} b_{j\alpha} \phi_\alpha + M \sum_{\beta \in \bar{\beta}} b_{j\beta} \phi_\beta \right] \tag{5.3}$$

where $\sum_{\alpha \in \bar{\alpha}} b_{j\alpha}^2 = 1$, $\sum_{\beta \in \bar{\beta}} b_{j\beta}^2 = 1$, λ is the normalization constant, and M denotes the mixing parameter between the symmetries $\bar{\alpha}$ and $\bar{\beta}$.

Putting (5.3) into the eigenvalue equation

$$H\psi_j = E_j \psi_j \tag{5.4}$$

where H is the total Hamiltonian, we get, after some simplification,

$$M^2 = \frac{H_{\bar{\alpha}\bar{\alpha}} - E_j}{H_{\bar{\beta}\bar{\beta}} - E_j} \tag{5.5}$$

where

$$H_{\bar{\alpha}\bar{\alpha}} = \sum_{\substack{\alpha \in \bar{\alpha} \\ \alpha' \in \bar{\alpha}}} b_{j\alpha} b_{j\alpha'} \langle \phi_\alpha | H | \phi_{\alpha'} \rangle. \tag{5.6}$$

In the mixing of the two symmetries, either $\bar{\alpha}$ is the dominant symmetry with $\bar{\beta}$ mixing into it due to symmetry-mixing terms in H or vice versa. Here we assume that $\bar{\alpha}$ is the dominant symmetry. In eqn (5.5) we replace E_j by the average energy

$$E = [\tfrac{1}{2}(H_{\bar{\alpha}\bar{\alpha}} + H_{\bar{\beta}\bar{\beta}}) - \{H_{\bar{\alpha}\bar{\alpha}} - H_{\bar{\beta}\bar{\beta}})^2 + 4H^2_{\bar{\alpha}\bar{\beta}}\}^{\frac{1}{2}}]. \tag{5.7}$$

Equation (5.5) can then be written

$$M^2 = \frac{(1+4\xi)^{\frac{1}{2}} - S}{(1+4\xi)^{\frac{1}{2}} + S} \tag{5.8}$$

where

$$\xi = \frac{H^2_{\bar{\alpha}\bar{\beta}}}{(H_{\bar{\alpha}\bar{\alpha}} - H_{\bar{\beta}\bar{\beta}})^2}, \tag{5.9a}$$

$$S = \frac{H_{\bar{\beta}\bar{\beta}} - H_{\bar{\alpha}\bar{\alpha}}}{|H_{\bar{\beta}\bar{\beta}} - H_{\bar{\alpha}\bar{\alpha}}|}. \tag{5.9b}$$

We can now express ξ and S in terms of the centroids $\varepsilon(\bar{\alpha})$, $\varepsilon(\bar{\beta})$, and the square of the partial width $\sigma^2(\bar{\alpha}, \bar{\beta})$ using the definitions (5.1) and (5.2). Thus

$$\xi = \frac{d^{-1}(\bar{\beta})\sigma^2(\bar{\alpha}, \bar{\beta})}{[\varepsilon(\bar{\beta}) - \varepsilon(\bar{\alpha})]^2}, \tag{5.10a}$$

$$S = \frac{\varepsilon(\bar{\beta}) - \varepsilon(\bar{\alpha})}{|\varepsilon(\bar{\beta}) - \varepsilon(\bar{\alpha})|}. \tag{5.10b}$$

Therefore the mixing parameter is given by eqn (5.8) with ξ and S given by eqns (5.10).

It is a simple matter to show that, if the partial width is much smaller than the difference in centroids, the expression for M^2 reduces to

$$M^2 = \frac{d^{-1}(\bar{\beta})\sigma^2(\bar{\alpha}, \bar{\beta})}{[\varepsilon(\bar{\beta}) - \varepsilon(\bar{\alpha})]^2}, \tag{5.11}$$

a result which can be derived directly using perturbation theory.

5.3 Examples of symmetry mixing

As mentioned in the introduction, in the nuclear shell model one is very often interested in the mixing of higher configurations with the dominant shell-model configurations. The formulation developed in § 5.2 can be used to obtain such information.

Consider the example of the nucleus ^{18}O which has two neutrons outside the ^{16}O core (Ullah 1971). The dominant configurations here are $2s_{\frac{1}{2}}$ and $1d_{\frac{5}{2}}$. Suppose we are interested in $J^\pi = 2^+$ and $T = 1$ states of ^{18}O built out of these two configurations; then the symmetry $\bar{\alpha}$ denotes all the states obtained using $2s_{\frac{1}{2}}$ and $1d_{\frac{5}{2}}$ configurations. Next we would like to see how the higher configuration $1d_{\frac{3}{2}}$ denoted by $\bar{\beta}$ gets mixed with $\bar{\alpha}$. For this purpose we shall need the centroids $\varepsilon(\bar{\alpha})$, $\varepsilon(\bar{\beta})$, and the square of the partial width $\sigma^2(\bar{\alpha}, \bar{\beta})$. A 40-MeV Rosenfeld interaction with single-particle energies, $\varepsilon_{2s_{\frac{1}{2}}} = 1.86$ MeV, $\varepsilon_{1d_{\frac{5}{2}}} = -1.10$ MeV, and $\varepsilon_{1d_{\frac{3}{2}}} = 3.39$ MeV, was used to obtain the matrix

$$\begin{array}{c} \\ \bar{\alpha}\begin{cases} 1 \\ 2 \end{cases} \\ \bar{\beta}\begin{cases} 1 \\ 2 \\ 3 \end{cases} \end{array} \begin{pmatrix} \overset{\bar{\alpha}}{1} & 2 & \overset{\bar{\beta}}{1} & 2 & 3 \\ -2.59226 & -0.43292 & -0.86455 & -0.91586 & 0.35348 \\ & -3.86047 & -0.33065 & -0.30612 & 1.24477 \\ & & 6.69613 & -0.52316 & 0.26997 \\ & & & 2.45011 & 0.24995 \\ & & & & 1.13770 \end{pmatrix}$$

Using the above matrix elements it turns out that the square of the mixing parameter M^2 given by eqns (5.8) and (5.10) has the value 0.0127. If we exactly diagonalize the above matrix and calculate the exact value of M^2 using eqn (5.3), it turns out to be 0.0172. Thus a knowledge of the centroids and partial width together with the expression for mixing in terms of these quantities gives a good estimate of the mixing of higher configurations.

Using different nuclear interactions Chakraborty et al. (1980) studied a very interesting problem of symmetry mixing, namely the validity of the spin–isospin SU(4) supermultiplet symmetry of Wigner for nuclei in 2s–1d shell and 1p–2f shell configurations. They also discussed recent techniques of evaluating centroids and partial widths. Their study showed that the Franzini–Radicatti mass ratio is insensitive to the mixing of SU(4) symmetry in nuclei in the 2s–1d and 1p–2f shells.

5.4 References

Chakraborty, M., Kota, V. K. B., and Parikh, J. C. (1980). *Physical Review Letters* **45**, 1073.
Chang, F. S. (1970). PhD thesis. University of Rochester, Rochester, New York.
French, J. B. (1967). Spectral distributions in nuclei. In *Nuclear structure* (ed. Anwar Hossain, Harun-Ar-Rashid, and Mizanul Islam). North Holland, Amsterdam. p. 85.
Preston, M. A. (1962). *Physics of the nucleus*. Addison-Wesley, Reading Massachusetts.
Preston, M. A. and Bhaduri, R. K. (1975). *Structure of the nucleus*. Addison-Wesley, Reading Massachusetts.
Ullah, N. (1971). *Nuclear Physics* **A164**, 658.
Wong, S. S. M. (1986). *Nuclear statistical spectroscopy*. Oxford University Press, Oxford.

6
ANTICORRELATION IN MANY-BODY SYSTEMS

6.1 Introduction

Much of the earlier work in atomic spectroscopy was based on the Hartree approximation. In this approximation the many-electron wave function is written as a product of independent single-electron wave functions. Later, in order to take into account the fermion nature of the electrons, Slater wrote the many-electron wave function as a determinant which is a totally antisymmetric wave function. We see that, whereas in the Hartree approximation the electrons were independent, this no longer holds when the wave function is written as a determinant. The single-electron wave function is now calculated using the Hartree–Fock self-consistent approximation.

It was Kutzelnigg and his collaborators (1968) who first showed the importance of the correlation coefficient for the electronic wave function. The two-particle correlation coefficient is defined as

$$\rho(\mathbf{r}_1, \mathbf{r}_2) = \frac{(\langle \mathbf{r}_1 \cdot \mathbf{r}_2 \rangle - \langle \mathbf{r}_1 \rangle \cdot \langle \mathbf{r}_2 \rangle)}{\prod_{k=1}^{2} (\langle r_k^2 \rangle - \langle \mathbf{r}_k \rangle \cdot \langle \mathbf{r}_k \rangle)^{\frac{1}{2}}}, \tag{6.1}$$

where $\mathbf{r}_1$ and $\mathbf{r}_2$ denote the position vectors of the two particles.

Later Banyard and Moore (1977) studied the effects of correlations on the one- and two-particle momentum distributions for H^-, He, and Li^+. Because of the difficulties in calculating correlation coefficients, only radial and angular correlation coefficients were calculated. The radial correlation coefficient in the momentum space were defined (Banyard and Moore 1977) as

$$\tau_p = \frac{\langle p_1 p_2 \rangle - \langle p_1 \rangle^2}{\langle p_i^2 \rangle - \langle p_i \rangle^2}, \tag{6.2a}$$

$$\tau_{\frac{1}{p}} = \frac{\langle p_1^{-1} \cdot p_2^{-1} \rangle - \langle p_1^{-1} \rangle^2}{\langle p_i^{-2} \rangle - \langle p_i^{-1} \rangle^2} \tag{6.2b}$$

while the angular correlation coefficients were given by

$$\tau_\gamma = \frac{\langle \mathbf{p}_1 \cdot \mathbf{p}_2 \rangle}{\langle p_i^2 \rangle}, \tag{6.3a}$$

6.2 THE CORRELATION COEFFICIENT

$$\tau_{\gamma'} = \frac{\left\langle \left(\frac{\mathbf{p}_1}{p_1^2}\right) \cdot \left(\frac{\mathbf{p}_2}{p_2^2}\right) \right\rangle}{\langle p_i^{-2} \rangle} \tag{6.3b}$$

where $i = 1$ or 2 and $\mathbf{p}_i$ is the momentum vector of the ith electron.

These studies seem to imply that the more negative the correlation coefficient the better would be the approximate many-body wave function.

In § 6.2 we shall describe an exactly solvable model to probe this point further while in § 6.3 we shall develop the concept of correlation between two parts of the total Hamiltonian. In § 6.4 we shall give examples of the use of correlation of two parts of the total Hamiltonian.

6.2 The correlation coefficient and energetically good wave functions

An exactly solvable model was used by King and Rothstein (1980) to show that more negative correlation does not necessarily imply a good wave function, i.e. a wave function having minimum energy as, for example, obtained by using a variational calculation. To show this they consider an exactly solvable Hamiltonian

$$H = -\tfrac{1}{2}\nabla_1^2 - \tfrac{1}{2}\nabla_2^2 - \tfrac{1}{2}k(r_1^2 + r_2^2) - \tfrac{1}{2}\alpha r_{12}^2 \tag{6.4}$$

where α is a parameter and $\mathbf{r}_{12} = \mathbf{r}_1 - \mathbf{r}_2$. We note from eqn (6.4) that for zero α the wave function would be a product of two single-electron wave functions, implying zero correlation.

We first transform the Hamiltonian to relative and centre-of-mass motion using the transformation

$$\mathbf{R} = \frac{1}{2^{\frac{1}{2}}}(\mathbf{r}_1 + \mathbf{r}_2), \tag{6.5a}$$

$$\mathbf{r} = \frac{1}{2^{\frac{1}{2}}}(\mathbf{r}_1 - \mathbf{r}_2). \tag{6.5b}$$

This gives

$$H = -\tfrac{1}{2}\nabla_R^2 - \tfrac{1}{2}\nabla_r^2 + \tfrac{1}{2}k(r^2 + R^2) - \alpha r^2. \tag{6.6}$$

The exact ground-state wave function can then be written as

$$\Psi^{(0)} = \psi(\mathbf{R})\phi(\mathbf{r}) \tag{6.7}$$

where

$$\Psi(\mathbf{R}) = \left(\frac{k^{\frac{1}{2}}}{\pi}\right)^{\frac{3}{4}} \exp(-\tfrac{1}{2}k^{\frac{1}{2}}R^2) \tag{6.8a}$$

and
$$\phi(\mathbf{r}) = \left(\frac{k_\alpha^{\frac{1}{2}}}{\pi}\right)^{\frac{3}{4}} \exp(-\tfrac{1}{2} k_\alpha^{\frac{1}{2}} r^2), \tag{6.8b}$$

k_α being given by
$$k_\alpha = (k - 2\alpha). \tag{6.8c}$$

The exact ground-state energy
$$E^{(0)} = \tfrac{3}{2}(k^{\frac{1}{2}} + k_\alpha^{\frac{1}{2}}). \tag{6.9a}$$

The exact correlation coefficient $\rho^{(0)}[\mathbf{r}_1, \mathbf{r}_2]$ from eqn (6.1) is given (King and Rothstein 1980) by

$$\rho^{(0)}[\mathbf{r}_1, \mathbf{r}_2] = \frac{(k_\alpha)^{\frac{1}{2}} - (k)^{\frac{1}{2}}}{(k_\alpha)^{\frac{1}{2}} + (k)^{\frac{1}{2}}}. \tag{6.9b}$$

It is obvious that the correlation coefficient will be zero if $k_\alpha = k$, i.e. $\alpha = 0$, as was remarked in the beginning of this section. The correlation coefficient will be -1 if $k_\alpha = 0$ or $\alpha = k/2$. Now consider an approximate wave function of the form

$$\bar\Psi = \psi(\mathbf{R})\phi^\beta(\mathbf{r}) \tag{6.10}$$

where $\psi(\mathbf{R})$ is given by expression (6.8a) and

$$\phi^\beta(\mathbf{r}) = \left(\frac{\beta^{\frac{1}{2}}}{\pi}\right)^{\frac{3}{4}} \exp(-\tfrac{1}{2}\beta^{\frac{1}{2}} r^2) \tag{6.11}$$

where β is a variational parameter.

For small values of α in eqn (6.4), the exact correlation coefficient is a small negative number. In eqn (6.11) we can choose the parameter β such that the correlation coefficient is close to -1. We find that, with this choice of β, the approximate ground-state energy $\bar E$ is $\simeq \tfrac{3}{2} k^{\frac{1}{2}}$ while the exact ground-state energy $E^{(0)}$ is $\simeq 3 k^{\frac{1}{2}}$. Thus $\bar\Psi$ is energetically a poor approximation to $\Psi^{(0)}$, despite the nearly perfect negative correlation.

6.3 Anticorrelation and the exact wave function

We mentioned in Chapters 2 and 3 that correlation between two operators of a many-body system is important in indicating the nature of the collectivity of a state. In this section we would like to show that there is complete anticorrelation between two parts of the total Hamiltonian if the exact eigenfunction of the total Hamiltonian is used (Ullah 1983). This information can be used to check how good an approximate eigenfunction is energetically. In some cases it may also be possible to obtain a trial form of the wave function which has this property.

6.3 ANTICORRELATION AND THE EXACT WAVE FUNCTION

Let H be the exact Hamiltonian of the system satisfying the exact eigenvalue equation

$$H\Psi = E\Psi. \tag{6.12}$$

We now split the total Hamiltonian H into two parts, H_0 and V, where V arises due to the correlations in the Hamiltonian. Then eqn (6.1) becomes

$$(H_0 + V)\Psi = E\Psi, \tag{6.13}$$

which immediately gives

$$E = \langle H_0 \rangle + \langle V \rangle \tag{6.14}$$

where $\langle \rangle$ denotes the expectation value of the enclosed operator with respect to Ψ.

We next multiply the left-hand side of eqn (6.13) by V to obtain

$$(VH_0 + V^2)\Psi = EV\Psi. \tag{6.15}$$

Rewriting eqn (6.15) in the form of a matrix element with respect to Ψ and using eqn (6.14)

$$\frac{\langle VH_0 \rangle - \langle V \rangle \langle H_0 \rangle}{\langle V^2 \rangle - \langle V \rangle^2} = -1 \tag{6.16}$$

By multiplying eqn (6.13) from the left by H_0 and carrying through the same steps we can also derive the relation

$$\frac{\langle H_0 V \rangle - \langle H_0 \rangle \langle V \rangle}{\langle H_0^2 \rangle - \langle H_0 \rangle^2} = -1. \tag{6.17}$$

Thus, we see from eqns (6.16) and (6.17) that the expectation value of the product of $(V - \langle V \rangle)$ with $(H_0 - \langle H_0 \rangle)$ divided by the mean-square deviation of H_0 or V using the exact wave function is -1.

If we calculate the value of the correlation coefficient

$$\rho = \frac{\langle VH_0 \rangle - \langle V \rangle \langle H_0 \rangle}{[(\langle H_0^2 \rangle - \langle H_0 \rangle^2)(\langle V^2 \rangle - \langle V \rangle^2)]^{\frac{1}{2}}}, \tag{6.18}$$

then, assuming H_0 to be Hermitian, ρ turns out to be -1 implying that there is complete anticorrelation between V and H_0 for the exact wave function.

A very general result of the type (6.16) and (6.17) can be derived for an arbitrary operator A belonging to the same Hilbert space as the Hamiltonian H by noting the expectation value $\langle A \rangle$ and $\langle H - A \rangle$ have the same variance. This arbitrariness is taken care of here by splitting the total Hamiltonian into two parts H_0 and V where the part V gives rise to correlations.

A convenient way of writing the anticorrelation identities is to express (Katriel and Ullah 1984) them using various variances as

$$\sigma_{H_0} = \sigma_V, \qquad \sigma_{H_0} = -\sigma_{H_0 V},$$
$$\sigma_V = -\sigma_{H_0 V}, \qquad \sigma_{H_0 V} = -(\alpha_{H_0}\sigma_V)^{\frac{1}{2}} \qquad (6.18)$$

where

$$\sigma_{H_0} = \langle H_0^2 \rangle_\psi - \langle H_0 \rangle_\psi^2,$$
$$\sigma_V = \langle V^2 \rangle_\psi - \langle V \rangle_\psi^2, \qquad (6.19)$$
$$\sigma_{H_0 V} = \langle H_0 V \rangle_\psi - \langle H_0 \rangle_\psi \langle V \rangle_\psi.$$

the notation $\langle \theta \rangle_\psi$ stands for $\langle \psi | \theta | \psi \rangle / \langle \psi | \psi \rangle$.

It can easily be shown (Katriel and Ullah 1984) that all four identities given by (6.18) are not independent; any two of them entail the other two. Further one can show that satisfaction of any two of the identities given by (6.18) is sufficient for the wave function involved to be an exact eigenfunction of the Hamiltonian.

One can also relate the variances given by (6.19) to the variance of H given by

$$\sigma_H = \langle H^2 \rangle_\psi - \langle H \rangle_\psi^2 \qquad (6.20)$$

so that

$$\sigma_H = \sigma_{H_0} + 2\sigma_{H_0 V} + \sigma_V. \qquad (6.21)$$

Using (6.21) we can discuss the minimization of the variance when a trial wave function ϕ is used for the exact wave function ψ.

6.4 Examples of the use of anticorrelation identities

The first example which we would like to consider is that of the exactly solvable model described in § 6.2. The total Hamiltonian H after the introduction of center-of-mass and relative coordinates is given by eqn (6.6) which is a sum of two oscillator Hamiltonians $H(\mathbf{R})$ and $H(\mathbf{r})$. Further, since in the trial wave function an exact eigenfunction of $H(\mathbf{R})$ is used and since there is complete symmetry between $x, y,$ and z directions, it is sufficient to consider the Hamiltonian

$$H = -\frac{1}{2}\frac{\partial^2}{\partial x^2} + Kx^2 + \lambda x^2 \qquad (6.22)$$

with

$$H_0 = -\frac{1}{2}\frac{\partial^2}{\partial x^2} + Kx^2 \qquad (6.23a)$$

6.4 THE USE OF ANTICORRELATION IDENTITIES

and
$$V = \lambda x^2. \tag{6.23b}$$

For the trial wave function ϕ we write
$$\phi = N \exp(-\tfrac{1}{2} \beta x^2) \tag{6.24}$$
where N is the normalization constant and β is the unknown parameter. It is now a simple matter to calculate the various matrix elements involving H_0 and V. This gives

$$\frac{\langle \phi | VH_0 | \phi \rangle - \langle \phi | V | \phi \rangle \langle \phi | H_0 | \phi \rangle}{\langle \phi | V^2 | \phi \rangle - \langle \phi | V | \phi \rangle^2} = \frac{2K - \beta^2}{2\lambda}. \tag{6.25}$$

According to eqn (6.16), ϕ will be the exact wave function provided the right-hand side of eqn (6.25) is -1. This gives us $\beta^2 = 2(K - \lambda)$. This is also the energetically best wave function for the Hamiltonian given by eqn (5.22) as can be confirmed easily by direct calculation. This shows that the concept of anticorrelation between two parts of the Hamiltonian can be used to obtain an energetically good wave function.

Our next example is based on the solvable model of Lipkin, Meshkov, and Glick (1965). This model has been very useful in studying various approximations for a many-body system. Using quasi-spin operators, the Hamiltonian
$$H = J_z + \tfrac{1}{2}(J_+^2 + J_-^2). \tag{6.26}$$
For this Hamiltonian, H_0 is taken to be J_z and $V = \tfrac{1}{2}(J_+^2 + J_-^2)$.

The trial ground-state wave function for a two-body system is written as
$$\phi = \cos \theta \, |1, -1\rangle + \sin \theta \, |1, 1\rangle, \tag{6.27}$$
where $|jm\rangle$ represent the eigenfunctions of J^2 and J_z and θ is the unknown parameter. Using angular momentun algebra it is easy to see that
$$\frac{\langle \phi | H_0 V | \phi \rangle - \langle \phi | H_0 | \phi \rangle \langle \phi | V | \phi \rangle}{\langle \phi | V^2 | \phi \rangle - \langle \phi | V | \phi \rangle^2} = \tan 2\theta. \tag{6.28}$$

Equation (6.16) tells us that ϕ will be the exact ground state if $\tan 2\theta = -1$ which gives $\theta = -\tfrac{1}{8}\pi$. By actual diagonalization of the Hamiltonian in eqn (6.26) we see that this is indeed the correct solution of the problem.

We next consider an example in which the trial function cannot be made an exact eigenfunction of the Hamiltonian (Katriel and Ullah 1984). We consider the Hamiltonian
$$H = p^2 + x^4, \tag{6.29}$$

where p is the one-dimensional momentum operator. In this case

$$H_o = p^2, \qquad V = x^4. \tag{6.30}$$

The trial eigenfunction ϕ is taken to be of the form

$$\phi = \left(\frac{2\alpha}{\pi}\right)^{\frac{1}{4}} \exp(-\alpha x^2) \tag{6.31}$$

with α as the unknown parameter. It is easy to see that for this case

$$\sigma_{H_0} = 2\alpha^2, \qquad \sigma_V = \frac{3}{8\alpha^4}, \qquad \sigma_{H_0 V} = -\frac{3}{4\alpha}$$

$$\langle H \rangle = \alpha + \frac{3}{16\alpha^2}, \qquad \alpha_H = 2\alpha^2 - \frac{3}{2\alpha} + \frac{3}{8\alpha^4}.$$

Minimizing $\langle H \rangle$, the value of α is $(3^{\frac{1}{3}}/2) = 0.721$. If one uses the anticorrelation identity $(\sigma_{H_0 V}/\sigma_{H_0}) = -1$, α is 0.721, while the identity $(\sigma_{H_0 V}/\sigma_V) = -1$ gives $(\sigma_{H_0 V}/\sigma_V)$ $\alpha = 0.794$. Thus, if ϕ cannot be made an exact eigenfunction of the Hamiltonian, one will get different values of the parameter using different anticorrelation identities. It is purely coincidental that the identity $(\sigma_{H_0 V}/\sigma_{H_0}) = -1$ gives the same value of α as that obtained using variational principles. This example illustrates the fact that, if the trial wave function ϕ cannot be made an exact eigenfunction, different anticorrelation identities will give different values of the unknown parameter in the trial wave function.

6.5 Concluding remarks

In this chapter we have considered simple one- and two-body Hamiltonians and by calculating various correlation coefficients have tried to get information about the goodness of trial wave functions. It is hoped that these ideas will further be generalized to an arbitrary number of fermions and made use of in deriving energetically-good many-fermion wave functions.

6.6 References

Banyard, K. E. and Moore, J. C. (1977). *Journal of Physics* **B10**, 2781.
Katriel, J. and Ullah, N. (1984). *Physical Review* **A29**, 2222.
King, F. W. and Rothstein, S. M. (1980). *Physical Review* **A21**, 1376.
Kutzelnigg, C. Del Re and Berthier, G. (1968). *Physical Review* **172**, 49.
Lipkin, H. J., Meshkov, N., and Glick, A. J. (1965). *Nuclear Physics* **62**, 188.
Ullah, N. (1983). *Physical Review* **A27**, 533.

7
APPROXIMATE EVALUATION OF PHYSICAL QUANTITIES

7.1 Introduction

It is often necessary to evaluate matrix elements of functions of operators, a simple example being the operator Ω of Chapter 3 which has the square root function $(1 + 4J^2)^{\frac{1}{2}}$, where J^2 is the operator which has eigenvalues $J(J+1)$. The exact evaluation of such an operator will involve the calculation of matrix elements of the form $\langle \Phi | J^{2n} | \Phi \rangle$ where $|\Phi\rangle$ is the deformed intrinsic wave function and n runs from 1 to some maximum value N. In the approximate evaluation we need to know the matrix elements of $\langle \Phi | J^{2n} | \Phi \rangle$ for only a few values of n.

Another example is the inverse operator which occurs in the theory of collective rotations developed by Peierls (Peierls and Urbano 1968). Here again we do not calculate the exact value of the matrix element of this operator which is quite difficult but we evaluate it approximately using the expectation value of the operator instead of the inverse operator.

Certain integrals also cannot be evaluated exactly, and we evaluate them approximately using similar kinds of approximations based on probability theory. In this chapter we shall give techniques for approximate evaluation of such quantities.

In § 7.2 we consider a simple integral to illustrate the approximation technique based on Taylor series expansion. In § 7.3 we describe a method using diagonalization in a truncated space. § 7.4 will describe the approximate evaluation of the matrix elements of functions of operators using low-order moments.

7.2 Approximate evaluation based on Taylor series expansion

Consider the integral

$$I = \int_{-\infty}^{\infty} \exp(-x^2)\cos(x^2)\, dx, \qquad (7.1)$$

which can easily be evaluated exactly as

$$I = \frac{\pi^{\frac{1}{2}}}{2^{\frac{3}{4}}} \cos \frac{\pi}{8}. \qquad (7.2)$$

44 APPROXIMATE EVALUATION OF PHYSICAL QUANTITIES

If we had a more complicated function than $\cos(x^2)$, it could be difficult to evaluate it exactly. In this case we could use the moment of x and expand the function around $\langle x^2 \rangle$ to calculate its approximate value. We illustrate the method using the function $\cos(x^2)$ given in integral (7.1).

Introducing the probability distribution

$$P(x) = \frac{1}{\pi^{\frac{1}{2}}} \exp(-x^2), \tag{7.3}$$

we rewrite I as

$$I = \pi^{\frac{1}{2}} \int_{-\infty}^{\infty} P(x)\cos(x^2)\, dx. \tag{7.4}$$

We now expand $\cos(x^2)$ around $\langle x^2 \rangle$ which gives the series

$$I = \pi^{\frac{1}{2}} \left[\int_{-\infty}^{\infty} P(x) \left\{ \cos\langle x^2 \rangle \left(1 - \frac{(x^2 - \langle x^2 \rangle)^2}{2} \cdots \right) \right. \right.$$
$$\left. \left. - \sin\langle x^2 \rangle \left((x^2 - \langle x^2 \rangle) - \frac{(x^2 - \langle x^2 \rangle)^3}{3!} \cdots \right) \right\} \right] dx. \tag{7.5}$$

It is obvious that the first term multiplying $\sin\langle x^2 \rangle$ is zero. Thus, I can be written approximately as

$$I = \pi^{\frac{1}{2}}[\cos\langle x^2 \rangle - \tfrac{1}{2}\cos\langle x^2 \rangle (\langle x^4 \rangle - \langle x^2 \rangle^2) + \ldots]. \tag{7.6}$$

The second term in the square brackets gives the fluctuation. If this is small, the integral can be approximated by

$$I = \pi^{\frac{1}{2}}\cos\langle x^2 \rangle. \tag{7.7}$$

From eqn (7.3) we can easily calculate $\langle x^2 \rangle$, $\langle x^4 \rangle$, etc. They are given by

$$\langle x^2 \rangle = \frac{1}{2}, \quad \langle x^4 \rangle = \frac{3}{4}. \tag{7.8}$$

Thus the fluctuation term in eqn (7.6) is 25 per cent of the first term. Putting the value of $\langle x^2 \rangle$ into eqn (7.7) we find that the first approximation to the value of I is $0.656\, \pi^{\frac{1}{2}}$ which should be compared to the exact value $0.777\, \pi^{\frac{1}{2}}$ given by eqn (7.2). Thus the error if we retain only the first term in expansion (7.6) is 13 per cent.

7.3 Approximation based on diagonalization in truncated space

Let us consider the evaluation of the matrix element of an inverse operator

$$M = \langle \phi_0 | \frac{1}{1+x^2} | \phi_0 \rangle \tag{7.9}$$

7.3 DIAGONALIZATION IN TRUNCATED SPACE

where

$$\phi_0 = (2\pi)^{-\frac{1}{4}} \exp(-\tfrac{1}{4} x^2) \tag{7.9a}$$

is the ground-state harmonic-oscillator wave function.

We have chosen this example of the inverse operator, instead of the more complicated directional operator used by Peierls and Urbano (1968), because the exact value of M can be found (Abramowitz and Stegun 1965) and can be used to test the accuracy of the approximation. The exact value of M is

$$M_{\text{exact}} = \left(\frac{\pi e}{2}\right)^{\frac{1}{2}} \text{erfc}\left(\frac{1}{2}\right) \tag{7.9b}$$

where erfc is the complementary error function (Abramowitz and Stegun 1965).

The basic idea of the approximation based on diagonalization in truncated space is to find a diagonal representation of the inverse operator with ϕ_0 as a member of this set.

Rather than working with the inverse operator $1/(1+x^2)$ we first rewrite M as

$$M = \langle \phi_0 | \frac{1}{1 + \langle x^2 \rangle + (x^2 - \langle x^2 \rangle)} | \phi_0 \rangle \tag{7.10}$$

where $\langle x^2 \rangle = \langle \phi_0 | x^2 | \phi_0 \rangle = 1$, using the wave function given by eqn (7.9a). Thus M can be written

$$M = \tfrac{1}{2} \langle \phi_0 | \frac{1}{1 + \frac{(x^2 - 1)}{2}} | \phi_x \rangle, \tag{7.11}$$

and we consider the diagonalization of the operator $\tfrac{1}{2}(x^2 - 1)$ in a truncated set. We note here that if we assume that the contribution of $x^2 - \langle x^2 \rangle$ in eqn (7.10) is small, our first approximation will give a value of M of 0.5 as in § 7.1.

For the next approximation we diagonalize $\tfrac{1}{2}(x^2 - 1)$ in the set consisting of ϕ_0 and ϕ_2 where

$$\phi_2 = (2\pi)^{-\frac{1}{4}} 2^{-\frac{1}{2}} (x^2 - 1) \exp(-\tfrac{1}{4} x^2). \tag{7.12}$$

It is easy to seen that the eigenvalue matrix is given by

$$\begin{pmatrix} 0 & \tfrac{1}{2} \\ \tfrac{1}{2} & 2 \end{pmatrix} \begin{pmatrix} x_0 \\ x_2 \end{pmatrix} = \varepsilon \begin{pmatrix} x_0 \\ x_2 \end{pmatrix} \tag{7.13}$$

where x_0 and x_2 are the expansion coefficients.

The eigenvalues are

$$\varepsilon_0 = 1 - (\tfrac{3}{2})^{\frac{1}{2}}, \qquad \varepsilon_2 = 1 + (\tfrac{3}{2})^{\frac{1}{2}} \tag{7.14a}$$

and the eigenvector is given by

$$\begin{pmatrix} x_0 \\ x_2 \end{pmatrix} = \begin{pmatrix} \dfrac{1}{(1+2\varepsilon^2)^{\frac{1}{2}}} \\ \dfrac{2}{(1+2\varepsilon^2)^{\frac{1}{2}}} \end{pmatrix} \quad (7.14b)$$

Denoting the diagonal representation by $|\chi_0\rangle$, $|\chi_2\rangle$, it is easy to see that

$$\begin{pmatrix} \chi_0 \\ \chi_2 \end{pmatrix} = \begin{pmatrix} \cos\Theta & \sin\Theta \\ -\sin\Theta & \cos\Theta \end{pmatrix} \begin{pmatrix} \phi_0 \\ \phi_2 \end{pmatrix} \quad (7.15)$$

where $\tan^2\Theta = -(\varepsilon_0/\varepsilon_2)$.

Since we are interested in the matrix elements of ϕ_0, we write using eqn (7.15)

$$\phi_0 = \chi_0 \cos\Theta - \chi_2 \sin\Theta. \quad (7.16)$$

Putting this in eqn (7.11) we get

$$M = \tfrac{1}{2}\left[\frac{\cos^2\Theta}{1+\varepsilon_0} + \frac{\sin^2\Theta}{1+\varepsilon_2}\right]. \quad (7.17)$$

Inserting the values of $\sin\Theta$, $\cos\Theta$, ε_0, and ε_2,

$$M = 0.60. \quad (7.17)$$

M_{exact} given by eqn (17.9b) is 0.65. Thus the error using ϕ_0 and ϕ_2 is 7 per cent.

For further improvement we can include ϕ_4 in the set and carry through the process again. However, the usefulness of the method decreases if we have to use too many wave functions in order to achieve a reasonable accuracy.

7.4 Method based on the use of lower-order moments

In variable moment-of-inertia models we encounter various functions of J^2; for example, in the Holmberg–Lipas (1968) model it is the square root function. Similarly, in the discussion of vibrations we need the matrix elements of the square root function $(1+4J^2)^{\frac{1}{2}}$. The exact evaluation of such functions is quite involved. Here we describe a method which makes use of the lower-order moments of J^2 to evaluate such matrix elements approximately.

Let us consider the evaluation of the matrix element

$$M = \langle\Phi|(1+4J^2)^{\frac{1}{2}}|\Phi\rangle \quad (7.19)$$

7.4 THE USE OF LOWER-ORDER MOMENTS

where Φ is a deformed intrinsic wave function which we assume consists of rotational wave functions Ψ_J.

It is easy to see that eqn (7.19) can be written

$$M = \int (1 + 4x)^{\frac{1}{2}} P(x) \, dx \qquad (7.20)$$

where

$$P(x) = \langle \Phi | \delta(x - J^2) | \Phi \rangle. \qquad (7.21)$$

The problem of evaluating the matrix element M has thus been changed to that of finding an approximate expression for $P(x)$.

We first note from eqn (7.21) that the variable x ranges from 0 to some maximum value depending on the highest J state contained in Φ. To simplify matters we take the upper limit as ∞. For the range of x between 0 and ∞, $P(x)$ can be approximated by an exponential multiplied by a series in x or Laguerre polynomials. We therefore write $P(x)$ as

$$P(x) = K \exp(-\lambda x), \qquad (7.22)$$

keeping only the exponential function where λ and K are constants determined by the normalization condition

$$\int_0^\infty P(x) \, dx = 1 \qquad (7.23a)$$

and the first moment of x

$$\int_0^\infty x P(x) \, dx = \langle x \rangle = \langle J^2 \rangle. \qquad (7.23b)$$

Thus $P(x)$ is approximately

$$P(x) = \langle J^2 \rangle^{-1} \exp\left(-\frac{x}{\langle J^2 \rangle}\right). \qquad (7.24)$$

We can now calculate other moments of x, such as $\langle x^2 \rangle = \langle J^4 \rangle$. From eqn (7.24) we find

$$\langle J^4 \rangle = 2 \langle J^2 \rangle^2. \qquad (7.25)$$

From Chapter 2, we know that, for the rotational problem, $\langle J^4 \rangle$ is approximately $2\langle J^2 \rangle^2$; thus the approximate form of $P(x)$ given by eqn (7.24) is fairly good. It should be noted here that, had we been discussing the collective vibrational problem, then (7.24) would not be such a good approximation as the ratio of $\langle J^4 \rangle / \langle J^2 \rangle^2$ is much larger than 2 for the vibrational problem. In this case one has to multiply the exponential function by a factor of the form $(1 + ax)$ where a is then fixed by the second moment of x.

Using the value of $P(x)$ given by eqn (7.24), we can write M using eqn (7.20) as

$$M = \langle J^2 \rangle^{-1} \int_0^\infty (1+4x)^{\frac{1}{2}} \exp\left(-\frac{x}{\langle J^2 \rangle}\right).$$

This can be integrated by a simple substitution to give

$$M = 1 + \left(\frac{\pi}{\lambda}\right)^{\frac{1}{2}} \exp\left(\frac{\lambda}{4}\right) \mathrm{erfc}\left(\frac{\lambda^{\frac{1}{2}}}{2}\right) \tag{7.26}$$

where $\lambda = 1/\langle J^2 \rangle$ and erfc is the complementary error function (Abramowitz and Stegun 1965).

Let us now use this approximation to calculate the value of $\langle \Phi | (1+4J^2)^{\frac{1}{2}} | \Phi \rangle$ where $|\Phi\rangle$ is the intrinsic deformed H–F wave function generated by Ripka (1968) for the nucleus ^{12}C. $\langle J^2 \rangle$ for this case is 7.24. Using this value of $\langle J^2 \rangle$ in eqn (7.26) we find that the approximate value of $\langle \Phi | (1+4J^2)^{\frac{1}{2}} | \Phi \rangle$ is 5.7 which is fairly close to its exact value of 4.8.

We now consider one more example which gives an approximation commonly used in the rotational case.

Consider the approximate evaluation of the matrix element

$$M = \langle \Phi | \exp(-i\beta J_y) | \Phi \rangle. \tag{7.27}$$

As in the earlier example we write

$$M = \int \exp(-i\beta x) P(x) \, dx \tag{7.28}$$

where $P(x)$ is given by

$$P(x) = \langle \Phi | \delta(x - J_y) | \Phi \rangle. \tag{7.29}$$

Since the range of x is now from $-\infty$ to ∞ we approximate $P(x)$ by a Gaussian

$$P(x) = K \exp(-ax^2). \tag{7.30}$$

The constants a and K can be easily found

$$a = (2\langle J_y^2 \rangle)^{-1}, \qquad K = (2\pi \langle J_y^2 \rangle)^{-\frac{1}{2}}. \tag{7.31}$$

Using eqns (7.28), (7.30), and (7.31) and integrating over x we easily find

$$M = \exp(-\tfrac{1}{2} \beta^2 \langle J_y^2 \rangle). \tag{7.32}$$

This approximation was checked by numerical calculations on angular momentum projections from an intrinsic deformed state. Thus we have shown how various approximations can be found using only the lower-order moments of angular momentum operators.

7.5 References

Abramowitz, M. and Stegun, A. (1965). *Handbook of mathematical functions.* Dover, New York.
Holmberg, P. and Lipas, P. O. (1968). *Nuclear Physics* **A117,** 552.
Peierls, R. E. and Urbano, J. H. (1968). *Proceedings of the Physical Society* **1,** 1.
Ripka, G. (1968). In *Advances in nuclear physics* (ed. M. Baranger and E. Vogt), Vol. 1, p. 183. Plenum Press, New York.

8
CORRELATIONS BETWEEN THE PARAMETERS OF THE S-MATRIX

8.1 Introduction

It is well well known that, when a beam of slow neutrons hits a target of a heavy nucleus such as ^{238}U we observe a number of narrow resonances in the total cross-section. It was Bohr who proposed a model of the compound nucleus to explain these observed resonances, according to which an incoming neutron and the target nucleus form a compound nucleus which then decays into many channels. The resonances which one observes are the quasi-stationary states above the neutron binding energy. It is obvious that such highly excited states will be extremely difficult to calculate in terms of the usual shell model picture of the nucleus. Because of the very large number of resonances, we do not calculate the parameters of each individual resonance but study the average properties of the positions of the compound nucleus levels and their widths. The matrix ensembles were originally introduced by Wigner (1957) to study these average properties. The most well known quantity is the distribution of the spacing of resonances which was derived by Wigner himself and is called the Wigner distribution.

It is the parameters of the R-matrix (Mahaux and Weidenmüller 1969; Lane and Thomas 1958) which are directly related to the eigenvalues and eigenvectors of a many-body Hamiltonian. Once the statistical properties of the R-matrix parameters are known, we can calculate the statistical properties of the S-matrix parameters by using the relations which connect S-matrix parameters with those of the R-matrix (Mahaux and Weidenmüller 1969; Lane and Thomas 1958).

We give a brief description of the R-matrix and its connection with the S-matrix in § 8.2. In § 8.3 we give the statistical distribution of R-matrix parameters. The statistical properties of the S-matrix parameters are discussed in § 8.4. § 8.5 considers the average value of the scattering matrix and the fluctuation of cross-sections. § 8.6 consists of some general remarks and an account of recent developments in this area.

8.2 The relation between the S- and R-matrices

As mentioned in the introduction the various cross-sections, such as the elastic scattering cross-section, total cross-section, etc., are written as a

8.2 THE S- AND R-MATRICES

function of the matrix elements of the scattering matrix S, while the statistical properties of the Hamiltonian are given in terms of the R-matrix. We therefore need a relation between the S- and R-matrices.

The essential idea of the R-matrix theory (Lane and Thomas 1958) of nuclear reactions is that the configuration space is divided into: (1) an internal space where all the particles interact and where the internal wave function X_λ obeys the eigenvalue equation

$$HX_\lambda = E_\lambda X_\lambda$$

where λ denotes the quantum numbers J and M and any other additional labels and H is the total Hamiltonian; and (2) an external space which is characterized by channels denoted by c. The channel index c stands for the channel spin, its component, and the relative motion of the two nuclei. We now define the reduced width amplitude

$$\gamma_{\lambda c} = \left(\frac{\hbar^2}{2M_c a_c}\right)^{\frac{1}{2}} \int \phi_c^* X_\lambda \, dS \tag{8.2}$$

where M_c is the reduced mass in channel c, a_c the channel radius, and dS is an element of channel surface. We shall give here only the important relevant quantities of R-matrix theory. For derivation and other details read the review article on R-matrix theory by Lane and Thomas (1958). In fact, for statistical distributions of many of the parameters of R-matrix theory, detailed knowledge of such constants as a_c and M_c separately is not needed as they are eventually replaced by some average value of $\gamma_{\lambda c}$. They are needed for making microscopic calculations from a given two-body potential.

The most important relation in R-matrix theory is

$$R_{cc'} = \sum_\lambda \frac{\gamma_{\lambda c} \gamma_{\lambda c'}}{E_\lambda - E}. \tag{8.3}$$

Since the $\gamma_{\lambda c}$s are real, the elements of the R-matrix are also real. Essentially the R-matrix gives the value of the logarithmic derivative of the internal wave function at the channel surface.

By matching the logarithmic derivative of the external wave function with that of the internal wave function we can connect the scattering matrix S with the R-matrix given by eqn (8.3). A fairly long derivation (Lane and Thomas 1958, Preston 1962) gives

$$S = \Omega^2 + 2i\Omega P^{\frac{1}{2}}(1 - RL^0)^{-1} RP^{\frac{1}{2}}\Omega \tag{8.4}$$

where $\Omega = I^{\frac{1}{2}} O^{-\frac{1}{2}}$, $L^0 = L - B$, and $L = S + iP$ are diagonal matrices; I and O represent the value at the channel surface of incoming and outgoing waves; B is a diagonal matrix giving real boundary conditions;

and L is defined in terms of the logarithmic derivatives of the incoming and outgoing waves evaluated at the channel surface mutliplied by the channel radius.

Equations (8.3) and (8.4) simplify considerably if purely elastic scattering is considered. In this case the matrix R becomes a simple function

$$R = \sum_\lambda \frac{\gamma_\lambda^2}{E_\lambda - E}, \tag{8.5}$$

where the single subscript c has been dropped from $\gamma_{\lambda c}$. The relation between the scattering matrix S and the R-function for purely elastic scattering becomes

$$S = \frac{1 - L^*R}{1 - LR} \exp(-i\phi) \tag{8.6}$$

where ϕ is a phase shift and L is a logarithmic derivative quantity with the real part denoted by $\mathcal{S}$ and the imaginary part by $\mathcal{P}$ which is called the penetration factor (Lane and Thomas 1958). The zero of energy is chosen so that the shift factor $\mathcal{S} = 0$. Thus the scattering function S for the purely elastic scattering case can be written

$$S = \frac{1 + iPR}{1 - iPR} \exp(-i\phi). \tag{8.7}$$

We can now see how the average properties of the Hamiltonian H give the average properties of the parameters of the scattering matrix using eqns (8.1), (8.3), (8.5), and (8.7).

8.3 The statistical distribution of the R-matrix parameters

Expanding X_λ in terms of wave functions ψ_μ

$$X_\lambda = \sum_\mu a_{\mu\lambda} \psi_\mu \tag{8.8}$$

where $a_{\mu\lambda}$ are the expansion coefficients, the eigenvalue eqn (8.1) can be written as a matrix equation

$$\sum_\theta (H_v\mu - E_\lambda \delta_v\mu) a_{\mu\lambda} = 0. \tag{8.9}$$

It can be shown that, for systems which possess time-reversal invariance and rotational invariance (Mehta 1967) the matrix of the Hamiltonian is a real symmetric matrix. Following Wigner we assume that each matrix element of H has an independent Gaussian distribution,

8.3 STATISTICAL DISTRIBUTION

the variance of the off-diagonal elements being half the variance of diagonal elements. Thus the joint distribution of the matrix elements of H can be written

$$P(\{H_{\mu\nu}\}) = K \exp - (\operatorname{Tr} H^2/2\sigma^2) \qquad (8.10)$$

where K is the normalization constant and σ^2 is the variance of the diagonal elements

We can now write the joint distribution of the N eigenvalues E_λ and the eigenvector components $a_{\mu\nu}$ using the relation between $H_{\mu\nu}$, E_λ, and $a_{\mu\nu}$. The most important result here is that the distribution can be written as a product of the distribution of the eigenvalues and that of the eigenvector components. To see this we first note from eqn (8.10) that, because of the invariance of tr, $\operatorname{tr} H^2 = \sum_\mu E_\mu^2$. Next we show that the volume element $\prod_{i \leq j} dH_{ij}$ is a product of volume elements in the space of eigenvector components and in that of the eigenvalues multiplied by a factor $\prod_{i<j} |E_i - E_j|$.

We first show this relation for a two-dimenstional case by using the Jacobian of the transformation. For the two-dimensional case we can write eqn (8.9) connecting the Hamiltonian H with eigenvalues E_λ and eigenvector components $a_{\mu\lambda}$

$$\begin{pmatrix} H_{11} & H_{12} \\ H_{12} & H_{22} \end{pmatrix} = \begin{pmatrix} \cos\theta & \sin\theta \\ -\sin\theta & \cos\theta \end{pmatrix} \begin{pmatrix} E_1 & 0 \\ 0 & E_2 \end{pmatrix} \begin{pmatrix} \cos\theta & -\sin\theta \\ \sin\theta & \cos\theta \end{pmatrix}.$$

Here the eigenvector components have been expressed in terms of the orthogonal matrix

$$\begin{pmatrix} \cos\theta & \sin\theta \\ -\sin\theta & \cos\theta \end{pmatrix}.$$

We can now calculate the volume element $dH_{11}\, dH_{22}\, dH_{12}$ in terms of $dE_1\, dE_1\, dE_2\, d\theta$ by calculating the Jacobian of the transformation

$$\begin{vmatrix} \dfrac{\partial H_{11}}{\partial E_1} & \dfrac{\partial H_{11}}{\partial E_2} & \dfrac{\partial H_{11}}{\partial \theta} \\ \dfrac{\partial H_{12}}{\partial E_1} & \dfrac{\partial H_{12}}{\partial E_2} & \dfrac{\partial H_{12}}{\partial \theta} \\ \dfrac{\partial H_{22}}{\partial E_1} & \dfrac{\partial H_{22}}{\partial E_2} & \dfrac{\partial H_{22}}{\partial \theta} \end{vmatrix}$$

By carrying out direct differentiation it is easy to see that the absolute value of the above determinant is $|E_1 - E_2|$. Thus,

$$dH_{11}\, dH_{12}\, dH_{22} = |E_1 - E_2|\, dE_1\, dE_2\, d\theta$$

which shows that the volume element is a product of the volume elements in the spaces of eigenvalues and eigenvector components.

We could generalize this result to N dimensions but we shall give here a different proof based on the line element in the space of Hamiltonian matrices.

The line element $(dS)^2$ in the space of real symmetric Hamiltonian matrices is defined by

$$(dS)^2 = \text{tr}(dH)^2. \tag{8.11}$$

Now, if E is the diagonal matrix having diagonal matrix elements E_μ and T is the orthogonal matrix,

$$H = TE\tilde{T} \tag{8.12}$$

where

$$T\tilde{T} = I. \tag{8.13}$$

From eqn (8.12),

$$dH = dT\, E\tilde{T} + T\, dE\, \tilde{T} + TE\, d\tilde{T}. \tag{8.14}$$

Also from (8.13)

$$d\tilde{T} = -\tilde{T}\, dT\, \tilde{T}. \tag{8.15}$$

Thus dH can be written

$$dH = dT\, E\tilde{T} + T\, dE\, \tilde{T} - TE\tilde{T}\, dT\, \tilde{T}. \tag{8.16}$$

Putting this in eqn (8.11) and using the cyclic property of trace and eqn (8.13)

$$(dS)^2 = \text{tr}(dE)^2 - 2\,\text{tr}(dE\, E\tilde{T}\, dT) + 2\,\text{tr}(\tilde{T}\, dT\, E\, dE) \\ + 2\,\text{tr}(dT\, E\tilde{T}\, dT\, E\tilde{T} - dT\, E^2\tilde{T}\, dT\, \tilde{T}). \tag{8.17}$$

It can easily be shown that the second and third term cancel each other.

Introducing the matrix dB defined by

$$dB = \tilde{T}\, dT, \tag{8.18}$$

it can be shown, using eqn (8.13) that dB is a real antisymmetric matrix; therefore, it has $\tfrac{1}{2} N(N-1)$ independent elements. After a few algebraic manipulations, eqn (8.17) can be written

$$(dS)^2 = \sum_\mu (dE_\mu)^2 + 2\sum_{k<n} (E_k - E_n)^2 [(dB)_{kn}]^2. \tag{8.19}$$

Thus the volume element† in the space of eigenvalues and eigenvector

† The volume element $dV = (\det g_{\mu\nu})^{\frac{1}{2}} \prod_\mu dq_\mu$ where $\det g_{\mu\nu}$ is the determinant of the metric tensor.

8.3 STATISTICAL DISTRIBUTION

components can be written as

$$\prod_{\mu<\nu} |E_\mu - E_\nu| \prod_\mu dE_\mu \prod_{\mu<\nu} (dB)_{\mu\nu},$$

which is of the product form. The product form of the volume element provides great simplification as we can now discuss separately the distribution of eigenvalues and eigenvector components. Thus, by integrating out the eigenvector components we can write the joint eigenvalue distribution of the eigenvalues E_λ as

$$P(\{E_\lambda\}) = K \exp\left(-\frac{\Sigma E_\mu^2}{2\sigma^2}\right) \prod_{\mu<\nu} |E_\mu - E_\nu| \qquad (8.20)$$

where K is the normalization constant. This is known as the Wishart distribution (Mehta 1967; Porter 1965).

We can integrate out all the eigenvalues except one from eqn (8.20) and thus derive the single-eigenvalue probability density function, which is known as Wigner's semi-circular distribution. The derivation is quite long; its exact form was first given by Mehta and Gaudin (1960) as

$$P(E) = (\pi N)^{-1}(2N - E^2)^{\frac{1}{2}}, \quad E^2 \leq 2N \qquad (8.21a)$$

where we have taken $\sigma^2 = 1$ and written E for any E_λ.

Expression (8.20) can also be used to find the probability density function $P(X)$ of the spacing of two resonances. For large N,

$$P(X) = \frac{\pi}{2} X \exp\left(-\frac{\pi}{4} X^2\right) \qquad (8.21b)$$

where $X = S/D$, S being the spacing and D the average spacing.

From eqn (8.20) we can also calculate (Mehta 1967) the variance of E_λ^2 and the average of $E_\lambda E_\mu$ for $\lambda \neq \mu$ as

$$\langle E_\mu^2 \rangle = \tfrac{1}{2}(N+1), \qquad (8.22a)$$

$$\langle E_\lambda E_\mu \rangle = -\tfrac{1}{2}. \qquad (8.22b)$$

It is now a simple matter to calculate the correlation coefficient to two eigenvalues,

$$\rho_{E_\lambda, E_\mu} = \frac{\langle E_\lambda E_\mu \rangle - \langle E_\lambda \rangle^2}{[\langle E_\lambda^2 \rangle - \langle E_\lambda \rangle^2]}.$$

Using expressions (8.22),

$$\rho_{E_\lambda, E_\mu} = -\frac{1}{N+1}. \qquad (8.23)$$

Thus, the correlation coefficient of two eigenvalues is a small negative quantity. This is verified experimentally from slow-neutron data.

We next consider the distribution of eigenvector components. This distribution will be used to find the distribution of the reduced width amplitude.

The first step is to find a convenient way of writing the distribution of the matrix elements of the orthogonal matrix T. Consider a column vector of this matrix; the only constraint on the elements in a single column is the normalization condition which can be taken care of by using a δ-function. Using expansion (8.8) we first write the reduced-width amplitude $\gamma_{\lambda c}$ given by eqn (8.2) as

$$\gamma_{\lambda c} = \sum_{\mu} a_{\mu\lambda} J_{\mu c} \qquad (8.24)$$

where $J_{\mu c}$ is the overlap integral

$$J_{\mu c} = \left(\frac{\hbar^2}{2M_c a_c}\right)^{\frac{1}{2}} \int \phi_c^* \psi_\mu \, dS.$$

We consider a single channel c and levels $\lambda = 1, 2, \ldots, n$ where $n < N$, N being the dimension of the orthogonal matrix. Using the δ-function technique, the probability density function of $\{\gamma_\lambda\}$ can be written (Ullah 1967) as

$$P(\{\gamma_\lambda\}) = K \int \left[\prod_{\lambda=1}^{n} \delta\left(\gamma_\lambda - \sum_{\mu=1}^{N} a_{\mu\lambda} J_{\mu c}\right)\right]$$

$$\left[\prod_{\lambda=1}^{n} \delta\left(\sum_{\mu=1}^{N} a_{\mu\lambda}^2 - 1\right)\right] \left[\prod_{\lambda<\lambda'}^{n} \delta\left(\sum_{\mu=1}^{N} a_{\mu\lambda} a_{\mu\lambda'}\right)\right]$$

$$\sum_{\mu=1}^{N} \prod_{\lambda=1}^{n} da_{\mu\lambda} \qquad (8.25)$$

where we have dropped the subscript c as we are considering a single channel and K stands for the appropriate normalization constant.

To evaluate the integral in (8.25) we make an orthogonal transformation on the variables $a_{\mu\lambda}$,

$$a'_{\nu\lambda} = \sum_{\mu=1}^{N} C_{\nu\mu} a_{\mu\lambda} \qquad (8.26)$$

and choose

$$C_{1\mu} = J_{\mu c} \left(\sum_{\mu=1}^{N} J_{\mu c}^2\right)^{-\frac{1}{2}}, \qquad \mu = 1, \ldots, N. \qquad (8.27)$$

Since C is an orthogonal matrix, the scalar products and the volume element remain invariant. Using this transformation in eqn (8.25) and carrying out the integrations,

$$P(\{\gamma_\lambda\}) = \frac{\Gamma(\frac{1}{2}N)}{\Gamma[\frac{1}{2}(N-n)]} (\pi N \langle \gamma^2 \rangle)^{-\frac{1}{2}n}$$
$$\left[1 - \sum_{\lambda=1}^{n} \gamma_\lambda^2/N\langle \gamma^2\rangle\right]^{\frac{1}{2}(N-n-2)} \quad (8.28)$$

where

$$\langle \lambda^2 \rangle = \langle \gamma_\lambda^2 \rangle = \frac{1}{N} \sum_{\mu=1}^{N} J_{\mu c}^2.$$

If we take $n=1$ and $N\to\infty$, this gives the probability density function of the single reduced-width amplitude γ as

$$P(\gamma) = (2\pi\langle\gamma^2\rangle)^{-\frac{1}{2}} \exp\left(-\frac{\gamma^2}{2\langle\gamma^2\rangle}\right), \quad (8.29a)$$

which is Gaussian.

We can now derive the well known Porter–Thomas distribution (Porter 1965) of the width Γ from eqn (8.29a). Since, apart from a constant, Γ is just γ^2, eqn (8.29a) gives

$$P(\Gamma) = (2\pi\langle\Gamma\rangle)^{-\frac{1}{2}}\Gamma^{-\frac{1}{2}} \exp -\Gamma/2\langle\Gamma\rangle. \quad (8.29b)$$

8.4 Statistical properties of the parameters of the S-matrix

First consider elastic scattering. In this case we write the scattering function S using the pole resonance form as

$$S = S^0\left[1 - i\sum_\mu \frac{g_\mu^2}{E - Z_\mu}\right] \quad (8.30)$$

where S^0 represents a phase factor, $Z_\mu = \varepsilon_\mu - \frac{1}{2}i\Gamma_\mu$, and the g_μs are amplitudes which are complex in general. For simplicity we shall choose the boundary condition such that $\text{Re}[L^0(1 - R^0L^0)^{-1}] = 0$.

In the early days of neutron spectroscopy the resonances observed were well separated, that is $\langle\Gamma\rangle/D \ll 1$, where $\langle\Gamma\rangle$ is the average width of the resonances and D the average spacing. In this case writing

$$R = R_0 + \sum_{\lambda=1}^{N} \frac{\gamma_\lambda^2}{E_\lambda - E}, \quad (8.31)$$

we can show that $\varepsilon_\mu \simeq E_\mu$ and $\Gamma_\mu \simeq 2P\gamma_\lambda^2$, where P is the penetration factor. Thus the statistical properties of the S-matrix parameters ε_μ and Γ_μ are the same as those of the R-matrix parameters.

As the energy of the neutron beam increases, the resonances start interfering with each other and the parameters of the S-matrix become a function of the parameters of the R-matrix. The most difficult problem here is to deal with the unitarity condition (Ullah and Warke 1968)

$$S^*S = 1, \tag{8.32}$$

which implies relations among the parameters of S-matrix.

It is convenient to introduce new amplitudes defined by

$$X_\lambda = (2P)^{\frac{1}{2}}\gamma_\lambda. \tag{8.33}$$

Since P is assumed to be a constant, the statistical properties of X_λ are the same as those of γ_λ. Using eqns (8.5), (8.7), (8.30), (8.31), and (8.33), we can write the identity (Sandhya Devi and Ullah 1972)

$$\prod_\mu (E - Z_\mu) = \prod_\mu (E - E_\mu) + \frac{i}{2}\sum X_\mu^2 \prod_{\nu \neq \mu}(E - E_\nu). \tag{8.34}$$

In writing eqn (8.34), the effect of R_0 is taken into account by redefining X_λ which again does not change its statistical properties. Equating the coefficients of E^{N-1} on both sides,

$$-\sum Z_\mu = -\sum E_\mu + \frac{i}{2}\sum X_\mu^2. \tag{8.35}$$

Equating real and imaginary parts in expression (8.35),

$$\sum \varepsilon_\mu = \sum E_\mu, \tag{8.36a}$$

$$\sum \Gamma_\mu = \sum X_\mu^2. \tag{8.36b}$$

Expressions (8.36) show that the average values of the parmaeters of the S-function are the same as those of the R-function,

$$\langle \varepsilon_\mu \rangle = \langle E_\mu \rangle = 0, \tag{8.37a}$$

$$\langle \Gamma_\mu \rangle = \langle X_\mu^2 \rangle. \tag{8.37b}$$

Further equating the coefficients of E^{N-2} in identity (8.34)

$$\sum_{\mu<\nu} Z_\mu Z_\nu = \sum_{\mu<\nu} E_\mu E_\nu - \frac{i}{2}\sum_\mu X_\mu^2 \left(\sum_{\nu \neq \mu} E_\nu\right). \tag{8.38}$$

Separating eqn (8.38) into its real and imaginary parts,

$$\sum_{\mu<\nu} \varepsilon_\mu \varepsilon_\nu - \tfrac{1}{4}\sum_{\mu<\nu} \Gamma_\mu \Gamma_\nu = \sum_{\nu<\nu} E_\mu E_\nu, \tag{8.39a}$$

$$\sum_{\mu<\nu} (\varepsilon_\mu \Gamma_\nu + \varepsilon_\nu \Gamma_\mu) = \sum_\mu x_\mu^2 \left(\sum_{\nu \neq \mu} E_\nu\right). \tag{8.39b}$$

8.4 PARAMETERS OF THE S-MATRIX

Taking the averages of eqns (8.39a) and (8.39b),

$$\langle \varepsilon_\mu \varepsilon_\nu \rangle - \tfrac{1}{4} \langle \Gamma_\mu \Gamma_\nu \rangle = \langle E_\nu E_\nu \rangle, \quad (8.40a)$$

$$2\langle \varepsilon_\mu \Gamma_\nu \rangle = \langle X_\mu^2 \rangle \langle E_\mu \rangle. \quad (8.40b)$$

It should be noted here that the right-hand side of (8.40b) is obtained because E_μ and X_μ of the R-function are independently distributed. On the other hand, the parameters of the S-function ε_μ and Γ_μ do not have an independent distribution. However, as can be seen by writing the correlation coefficient between ε_μ and Γ_ν ($\mu \neq \nu$)

$$\rho_{\varepsilon_\mu, \Gamma_\nu} = \frac{\langle \varepsilon_\mu \Gamma_\mu \rangle - \langle \varepsilon_\mu \rangle \langle \Gamma_\mu \rangle}{[(\langle \varepsilon_\mu^2 \rangle - \langle \varepsilon_\mu \rangle^2)(\langle \Gamma_\mu^2 \rangle - \langle \Gamma_\mu \rangle^2)]^{\frac{1}{2}}}, \quad (8.41)$$

its value is zero because of relations (8.37a) and (8.40b). Thus there is no correlation between the parameters ε_μ and Γ_ν of the S-function. This will not be true for higher powers of ε_μ and Γ_ν as can be seen by equating other powers of E^k in the identity (8.34).

Further multiplying eqns (8.36a) and (8.36b) and taking the averages, we can show that

$$\langle \varepsilon_\mu \Gamma_\mu \rangle = 0, \quad (8.41)$$

which implies

$$\rho_{\varepsilon_\mu, \Gamma_\mu} = 0. \quad (8.42)$$

Other interesting correlations which one can derive for the S-function parameters are the width–width correlation coefficient and the correlation coefficient of $\varepsilon_\mu, \varepsilon_\lambda$.

For the width–width correlation we take the ensemble average of the square of eqn (8.36b). This gives

$$\langle \Gamma_\mu^2 \rangle + (N-1) \langle \Gamma_\mu \Gamma_\lambda \rangle_{\mu \neq \lambda} = \langle X_\mu^4 \rangle + (N-1) \langle X_\mu^2 X_\lambda^2 \rangle_{\mu \neq \lambda}. \quad (8.43)$$

From the distribution given by eqn (8.28), it can be shown that the ensemble averages of the real amplitudes X_μ are given by

$$\langle X_\mu^4 \rangle = \frac{3N}{N+2} \langle X_\mu^2 \rangle^2, \quad (8.44a)$$

$$\langle X_\mu^2 X_\lambda^2 \rangle_{\mu \neq \lambda} = \frac{N}{N+2} \langle X_\mu^2 \rangle^2. \quad (8.44b)$$

Expressions (8.43) and (8.44) give

$$\rho_{\Gamma_\mu, \Gamma_\lambda} = -\frac{1}{N-1}. \quad (8.45)$$

Thus the width–width correlation for the S-function is the same as that found for the R-function. It is weakly negatively correlated.

The calculation of the correlation coefficient of ε_μ and ε_λ is fairly long. We can show that it can be written as

$$\rho_{\varepsilon_\mu, \varepsilon_\lambda} = -\frac{1}{N+1} - \frac{2}{2(N-2)(N+1)} \frac{\langle \Gamma_\mu^2 \rangle - \langle \Gamma_\mu \rangle^2}{\langle \varepsilon_\mu^2 \rangle} + \frac{1}{2(N+1)} \frac{\langle \Gamma_\mu \rangle^2}{\langle \varepsilon_\mu^2 \rangle}. \quad (8.46)$$

We see from (8.46) that, if the resonances are isolated, the last two terms become small and

$$\rho_{\varepsilon_\mu, \varepsilon_\lambda} = -\frac{1}{N+1},$$

which is just the correlation coefficient of two real eigenvalues E_μ and E_λ as it should be.

So far we have not said anything about the complex amplitudes g_μ which enter into the collision function given by eqn (8.30). It can be shown (Ullah and Warke 1968) that, because of the unitarity condition given by eqn (8.32),

$$g_\mu^2 = \Gamma_\mu \prod_{\nu \neq \mu} (Z_\mu - Z_\nu^*)(Z_\mu - Z_\nu)^{-1} \quad (8.47)$$

where $Z_\mu = \varepsilon_\mu - \tfrac{1}{2} i\Gamma_\mu$.

An immediate consequence of eqn (8.47) is

$$\sum_{\mu=1}^{N} g_\mu^2 = \sum_{\mu=1}^{N} \Gamma_\mu. \quad (8.48)$$

Thus the average value of g_μ^2 is real and

$$\langle g_\mu^2 \rangle = \langle \Gamma_\mu \rangle. \quad (8.49)$$

We could ask what are the effects of the unitarity constraint on the statistical properties of the collision function. For this purpose we can calculate the mean-square deviation of the width Γ_μ. After a farily long calculation (Sandhya Devi and Ullah 1972) we find that

$$\frac{\langle \Gamma_\mu^2 \rangle - \langle \Gamma_\mu \rangle^2}{\langle \Gamma_\mu \rangle^2} = \frac{3N}{N+1} \left(\frac{1 + \alpha N(3N+32)/(N+4)(N+6)}{1 + 3\alpha N/(N+2)} \right) - 1 \quad (8.50)$$

where the parameter $\alpha = \pi^2 \langle X_\mu^2 \rangle^2 / 16 d^2$, d being the spacing of the poles of the R-function.

The mean-square deviation of the square of the amplitude X_μ using (8.28) is

$$\frac{(\langle X_\mu^4 \rangle - \langle X_\mu^2 \rangle^2)}{\langle X_\mu^2 \rangle^2} = 2 \frac{(N-1)}{N+2}. \quad (8.51)$$

8.4 PARAMETERS OF THE S-MATRIX

A comparison of eqns (8.50) and (8.51) shows that the width Γ_μ has a larger mean-square deviation than the square of the amplitude X_μ.

So far we have discussed only the purely elastic scattering case; we shall now discuss the multichannel problem. For this problem we need to know the multichannel distribution of the reduced width amplitudes $\{\gamma_{\lambda c}\}$. We consider a single level and m channels. Since we are considering a single level, we drop the label λ. Using eqn (8.24) we can write (Ullah 1967) the multichannel probability density function $P(\{\gamma_c\})$ as

$$P(\{\gamma_c\}) = K \int \prod_{j=1}^{m} \delta\left(\gamma_{c_j} - \sum_{\mu=1}^{N} a_\mu J_{\mu c_j}\right) \delta\left(\sum_{\mu=1}^{N} a_\mu^2 - 1\right) \prod_{\mu=1}^{N} da_\mu \quad (8.52)$$

where c_j denotes a particular channel and K the normalization constant.

By taking proper linear combinations of the quantities $J_{\mu c_j}$

$$J'_{\mu c_i} = \sum_{j=1}^{m} T_{ij} J_{\mu c_j}, \quad (8.53)$$

we can construct $J'_{\mu c_i}$ which form m orthonormal vectors in N-dimensional space. Expression (8.52) then becomes

$$P(\{\gamma_c\}) = K \int \prod_{j=1}^{m} \delta(\gamma_{cj} - \sum_{\mu=1, k=1}^{Nm} a_\mu (T^{-1})_{jk} J'_{\mu c_k}) \delta\left(\sum_{\mu=1}^{N} a_\mu^2 - 1\right) \prod_{\mu=1}^{N} da_\mu. \quad (8.54)$$

By making an orthogonal transformation on the variables a_μ we can carry through the integrations over a_μ as was done earlier. This gives

$$P(\{\gamma_c\}) = \frac{\Gamma(\frac{1}{2}N)|T|}{\Gamma[\frac{1}{2}(N-m)]\pi^{\frac{1}{2}m}} [1 - (\gamma, \tilde{T}T\gamma)]^{\frac{1}{2}(N-m-2)} \quad (8.55)$$

where $|T|$ is the determinant of the matrix T. Introducing the covariance matrix Σ (Anderson 1958),

$$\Sigma = \langle \gamma \tilde{\gamma} \rangle = N^{-1}(\tilde{T}T)^{-1},$$

eqn (8.55) becomes

$$P(\{\gamma_c\}) = \frac{\Gamma(\frac{1}{2}N)}{\Gamma[\frac{1}{2}(N-m)](\pi N)^{\frac{1}{2}m}|\Sigma|^{\frac{1}{2}}} \left[1 - \frac{1}{N}(\gamma, \Sigma^{-1}\gamma)\right]^{\frac{1}{2}(N-m-2)}. \quad (8.56)$$

As before we get a simple distribution if the number of channels is small, say $m = 2$, by writing the power as exponential. The two-channel distribution can be written

$$P(\gamma_{c1}, \gamma_{c2}) = (2\pi |\Sigma|^{\frac{1}{2}})^{-1} \exp -\tfrac{1}{2}(\gamma, \Sigma^{-1}\gamma). \quad (8.57)$$

Thus we see that, unlike the multilevel distribution, the multichannel distribution of the reduced-width amplitudes is a correlated Gaussian dsitribution.

For the multichannel case the pole resonance form of the scattering matrix S is given by

$$S = V\left(1 - i \sum_{\lambda=1}^{N} \frac{g_\lambda \times g_\lambda}{E - Z_\lambda}\right) V \qquad (8.58)$$

where V is a unitary and symmetric matrix which gives rise to background scattering, $g_\lambda \times g_\lambda$ denotes the matrix formed from the complex amplitudes g_λ and $Z_\lambda = \varepsilon_\lambda - \frac{1}{2} i \Gamma_\lambda$. $(g_\lambda \times g_\lambda)$ is a matrix in channel space, the cc' element being $(g_{\lambda c} g_{\lambda c'})$.

The relation between Z_λ and the parameters of the R-matrix is now given by the identity

$$\prod_{\lambda=1}^{N} (E - Z_\lambda) = \det(A^{-1}) \qquad (8.59)$$

where

$$(A^{-1})_{\lambda\mu} = (E - E_\lambda)\delta_{\lambda\mu} + \frac{i}{2}(X_\lambda, X_\mu) \qquad (8.60)$$

where E_λ and X_λ are real eigenvalues and real amplitudes of the R-matrix and (X_λ, X_μ) denotes the scalar product of X_λ and X_μ in the channel space.

The identity (8.60) gives the relations among the parameters of the S- and R-matrices

$$\sum_{\mu=1}^{N} \varepsilon_\mu = \sum_{\mu=1}^{N} E_\mu, \qquad (8.61a)$$

$$\sum_{\mu=1}^{N} \Gamma_\mu = \sum_{\mu=1}^{N} \sum_{c=1}^{m} X_{\mu c}^2, \qquad (8.61b)$$

$$\sum_{\mu<\lambda}^{N} (\varepsilon_\mu \varepsilon_\lambda - \tfrac{1}{4} \Gamma_\mu \Gamma_\lambda) = \sum_{\mu<\lambda} E_\mu E_\lambda - \tfrac{1}{4} \sum_{\mu<\lambda} (X_{\mu c} X_{\lambda c'} - X_{\mu c'} X_{\lambda c})^2, \; c<c', \qquad (8.61c)$$

$$\sum_{\mu<\lambda}^{N} (\varepsilon_\mu \Gamma_\lambda + \varepsilon_\lambda \Gamma_\mu) = \sum_{\mu, c} X_{\mu c}^2 \left(\sum_{\lambda \neq \mu} E_\lambda\right). \qquad (8.61d)$$

We now calculate the correlation coefficient between two widths. For this the ensemble averages of eqn (8.61b) and its square. This gives

$$\langle \Gamma_\mu \rangle = \sum_{c=1}^{m} \langle x_{\mu c}^2 \rangle, \qquad (8.62a)$$

8.5 AVERAGE VALUE OF THE SCATTERING MATRIX

$$\langle \Gamma_\mu^2 \rangle + (N-1)\langle \Gamma_\mu \Gamma_\lambda \rangle_{\mu \neq \lambda} = \sum_{c=1}^{m} [\langle X_{\mu c}^4 \rangle + (N-1)\langle X_{\mu c}^2 X_{\lambda c}^2 \rangle_{\mu \neq \lambda}]$$

$$+ \sum_{c \neq c'} [\langle X_{\mu c}^2 X_{\mu c'}^2 \rangle + (N-1)\langle X_{\nu c}^2 X_{\lambda c'}^2 \rangle_{\mu \neq \lambda}]. \quad (9.62b)$$

The ensemble averages of the real amplitudes in the first square bracket were already given when we considered purely elastic scattering. The ensemble average of the first term in the second square bracket can be calculated using the distribution (8.56), giving

$$\langle X_{\mu c}^2 X_{\mu c'}^2 \rangle_{c \neq c'} = \frac{N}{N+2} \langle X_{\mu c}^2 \rangle \langle X_{\mu c'}^2 \rangle (1 + 2\rho_{X_{\mu c}, X_{\mu c'}}^2) \quad (8.63)$$

where $\rho_{X_{\mu c}, X_{\mu c'}}$ is the reduced-width amplitude correlation coefficient.

After some algebra, the last term in the second square bracket can be written

$$\langle X_{\mu c}^2 X_{\mu c'}^2 \rangle_{\substack{\mu \neq \lambda \\ c \neq c'}} = \frac{N(N+1)}{(N-1)(N+2)} \langle X_{\mu c}^2 \rangle \langle X_{\mu c'}^2 \rangle (1 - 2\rho_{X_{\mu c}, X_{\mu c'}}^2). \quad (8.64)$$

Using eqns (8.62)–(8.64) we find that the width–width correlation coefficient for the multichannel case is given by

$$\rho_{\Gamma_\mu, \Gamma_\mu} = -\frac{1}{N-1}. \quad (8.65)$$

Thus the correlation coefficient between two total widths is the same as that found earlier in the elastic scattering case.

We can also calculate the correlation coefficient between two ε_μs. It can be shown that it is given by

$$\rho_{\varepsilon_\mu, \varepsilon_\lambda}^{(\text{inelastic})} = \rho_{\varepsilon_\mu, \varepsilon_\lambda}^{(\text{elastic})}$$

$$+ \frac{N}{2(N-1)(N+1)} \frac{1}{\langle \varepsilon_\mu^2 \rangle} \sum_{c \neq c'} \langle X_{\mu c}^2 \rangle \langle X_{\mu c'}^2 \rangle (\rho_{X_{\mu c}, X_{\mu c'}}^2 - 1) \quad (8.66)$$

where the superscripts, elastic and inelastic refer to the purely elastic scattering case and the multichannel case, respectively.

Since $\rho_{X_{\mu c}, X_{\mu c'}}^2 \leq 1$, the effect of open channels c, is to decrease the correlation coefficient of the two ε_μs.

8.5 Average value of the scattering matrix and the fluctuation of cross-sections

In the last two sections we studied the average properties of the parameters of the R- and S-matrices. Since various cross-sections, e.g.

the partial reaction cross-section $\sigma_{cc'}$ given (Lane and Thomas 1958) by

$$\sigma_{cc'} = \frac{\pi}{k_c^2} |S_{cc'}|^2 \tag{8.67}$$

where k_c is the wave number in channel c, are expressed in terms of the matrix elements of the scattering matrix, we can now study the average values of the various cross-sections themselves by evaluating the averages of the matrix elements of S. We deal with two kinds of averages: (1) averages over the energy E; and (2) averages over the parameters of S. In practice we first average over energy E using some energy weighting function. The two most commonly used energy weighting functions are: (1) the box weight function; and (2) the Lorentzian weight function. Mathematically it is much easier to use the Lorentizian form of energy resolution function

$$\rho(E, E_0) = \frac{I}{\pi} \frac{1}{(E - E_0)^2 + \frac{I^2}{4}} \tag{8.68}$$

where $\frac{1}{2} I$ is the half-width of the Lorentzian and E_0 is the centre of the Lorentzian. The simplest of the averages of the matrix elements of S is the average of the diagonal elements of S. They are also physically interesting because they are related to the complex optical model phase shifts. In addition, once $\langle S_{cc} \rangle$ is known, the transmission coefficient for channel c is also known

$$T_c = 1 - |\langle S_{cc} \rangle|^2. \tag{8.69}$$

We consider the case of isolated resonances, for which

$$S_{cc}(E) = 1 - i \sum_\lambda \frac{\Gamma_{\lambda c}}{E - E_\lambda + \frac{i}{2} \Gamma_\lambda}$$

where we have omitted an over-all phase shift factor.

Using eqn (8.66) the energy average of $S_{cc}(E)$ is

$$\langle S_{cc} \rangle = \int_{-\infty}^{\infty} S_{cc}(E) \rho(E, E_0) \, dE.$$

This integral can easily be evaluated by going over to complex energies Z and by noting that there is a single pole in the upper half plane at $Z = E_0 + \frac{1}{2} iI$. By the theory of residues,

$$\langle S_{cc} \rangle = 1 - i \sum_\lambda \frac{\Gamma_{\lambda c}}{(E_0 - E_\lambda) + \frac{i}{2}(I + \Gamma_\lambda)}. \tag{8.70}$$

8.5 AVERAGE VALUE OF THE SCATTERING MATRIX

To evaluate the sum over λ, first neglect Γ_λ in comparison with I and remove the average of $\langle \Gamma_{\lambda c} \rangle$ from the sum, giving

$$\langle S_{cc} \rangle = 1 - i \langle \Gamma_{\lambda c} \rangle \sum_\lambda \frac{1}{(E_0 - E_\lambda) + \dfrac{iI}{2}}. \tag{8.71}$$

The sum over λ is evaluated by replacing it by an integral. Assume D to be the average spacing and write

$$\sum_\lambda \frac{1}{(E_0 - E_\lambda) + \dfrac{iI}{2}} = \frac{1}{D} \int_{-\infty}^{\infty} \frac{dE_\lambda}{(E_0 - E_\lambda) + \dfrac{iI}{2}} \tag{8.72}$$

$$= -\frac{i\pi}{D}. \tag{8.73}$$

Thus,

$$\langle S_{cc} \rangle = 1 - \frac{\pi \langle \Gamma_{\lambda c} \rangle}{D}. \tag{8.74}$$

Before we proceed further, we remark that the sum in (8.72) can also be calculated using the distribution of single eigenvalues E_λ giving

$$\sum_\lambda \frac{1}{(E_0 - E_\lambda) + \dfrac{iI}{2}} = \int dE_\lambda \rho(E_\lambda) \frac{1}{(E_0 - E_\lambda) + \dfrac{iI}{2}} \tag{8.75}$$

where

$$\rho(E_\lambda) = \frac{\pi}{2ND^2} \left(\frac{4N^2 D^2}{\pi^2} - E_\lambda^2 \right)^{\frac{1}{2}}. \tag{8.76}$$

N is the number of resonances, D the average spacing and $\langle E_\lambda \rangle = E_0 = 0$.

Carrying out the integrations,

$$\sum_\lambda \frac{1}{(E_0 - E_\lambda) + \dfrac{iI}{2}} = -\frac{i\pi}{D} \left[\left(1 + \frac{\pi^2 I^2}{16 N^2 D^2} \right)^{\frac{1}{2}} - \frac{\pi I}{4ND} \right]. \tag{8.77}$$

Since $I/ND \ll 1$, we get the same result as in eqn (8.73).

It is now a simple matter to write the transmission coefficient

$$T_c = \frac{2\pi \langle \Gamma_{\lambda c} \rangle}{D}. \tag{8.78}$$

Under the same assumptions we now derive an expression for the average reaction cross-section $\langle \sigma_{cc'} \rangle$. The matrix element $S_{cc'}$ is written

$$S_{cc'}(E) = -i \sum_\lambda \frac{(\Gamma_{\lambda c} \Gamma_{\lambda c'})^{\frac{1}{2}}}{E - E_\lambda + \frac{i}{2}\Gamma_\lambda}. \tag{8.79}$$

In calculating the energy averages of $|S_{cc'}(E)|^2$, we note that, in addition to the pole $Z = E_0 + \frac{1}{2}iI$, there will be other poles coming from $S^*_{cc'}(E)$ and $Z = \varepsilon_\nu + \frac{1}{2}i\Gamma_\nu$. This gives the energy average of $|S_{cc'}|^2$,

$$\langle |S_{cc'}|^2 \rangle = \sum_{\lambda, \lambda'} \frac{(\Gamma_{\lambda c}\Gamma_{\lambda c'}\Gamma_{\lambda' c}\Gamma_{\lambda' c'})^{\frac{1}{2}}}{\left[(E_0 - E_\lambda) + \frac{i}{2}(I + \Gamma_\lambda)\right]\left[(E_0 - E_{\lambda'}) + \frac{i}{2}(I - \Gamma_{\lambda'})\right]}$$

$$+ iI \sum_{\lambda, \lambda'} \frac{(\Gamma_{\lambda c}\Gamma_{\lambda c'}\Gamma_{\lambda' c}\Gamma_{\lambda' c'})^{\frac{1}{2}}}{(E_{\lambda'} - E_\lambda) + \frac{i}{2}(\Gamma_{\lambda'} + \Gamma_\lambda)} \frac{1}{\left[(E_{\lambda'} - E_0) + \frac{i}{2}\Gamma_{\lambda'}\right]^2 + \frac{I^2}{4}}. \tag{8.80}$$

As before we use the condition $I \gg \Gamma_\lambda$ and take the average of the widths outside the summation sign. Further, for $\lambda \neq \lambda'$, we assume that the average of two amplitudes $\gamma_{\lambda c}$ is zero. This gives

$$\langle |S_{cc'}|^2 \rangle = \langle \Gamma_{\lambda c} \rangle \langle \Gamma_{\lambda c'} \rangle \sum_\lambda \frac{1}{\left(E_0 - E_\lambda - i\frac{I}{2}\right)^2}$$

$$- \frac{\langle \Gamma_{\lambda c} \rangle \langle \Gamma_{\lambda c'} \rangle}{i \langle \Gamma_\lambda \rangle} \sum_\lambda \frac{1}{E_\lambda - E_0 + i\frac{I}{2}}$$

$$+ \frac{\langle \Gamma_{\lambda c} \rangle \langle \Gamma_{\lambda c'} \rangle}{i \langle \Gamma_\lambda \rangle} \sum_\lambda \frac{1}{E_\lambda - E_0 - i\frac{I}{2}}. \tag{8.81}$$

The summations over λ can now be carried out in the same way as already done for the average S_{cc}. The first term makes a zero contribution, while the last two terms make equal contribution, giving

$$\langle |S_{cc'}|^2 \rangle = 2\pi \frac{\langle \Gamma_{\lambda c} \rangle \langle \Gamma_{\lambda c'} \rangle}{\sum_{c''} \langle \Gamma_{\lambda c''} \rangle}. \tag{8.82}$$

Writing $\pi \langle \Gamma_{\lambda c} \rangle / D$ in terms of the transmission coefficient T_c,

$$\langle |S_{cc'}|^2 \rangle = \frac{T_c T_{c'}}{\sum_{c''} T_{c''}}. \tag{8.83}$$

8.5 AVERAGE VALUE OF THE SCATTERING MATRIX

Thus, the average reaction cross-section

$$\langle \sigma_{cc'} \rangle = \frac{\pi}{k_c^2} \frac{T_c T_{c'}}{\sum_{c''} T_{c''}}. \tag{8.84}$$

This is the Hauser–Feshbach expression for the average reaction cross-section.

As the excitation energy increases, the resonances begin to overlap. For overlapping resonances, Ericson (1963) calculated the variances of cross-sections and the energy correlation functions for the cross-sections at two different energies. We now briefly describe these results.

The starting point is the matrix element of the scattering matrix S, which Ericson (1963) wrote as

$$S_{cc'}(E) = S_{cc'}^{(p)} + i \sum_i \frac{a_i}{E - E_i} \tag{8.85}$$

where

$$E_i = \operatorname{Re} E_i - \tfrac{1}{2} i \Gamma_i \tag{8.86a}$$

and

$$a_i = \langle a \rangle + \delta a_i. \tag{8.86b}$$

Thus, $S_{cc'}$ can be split into two parts

$$S_{cc'} = \langle S_{cc'} \rangle + S_{cc'}^{\text{fl}}(E) \tag{8.87}$$

where

$$\langle S_{cc'} \rangle = S_{cc'}^{(p)} + \frac{\pi \langle a \rangle}{D} \tag{8.88}$$

and the fluctuating matrix element

$$S_{cc'}^{\text{fl}} = i \sum_i \frac{\delta a_i}{E - E_i}. \tag{8.89}$$

It is assumed that we are discussing the scattering of spinless particles from target nuclei having zero spin. The $\langle S_{cc'} \rangle$ part gives rise to the direct reaction cross-section $\sigma_{cc'}^{\text{DI}}$ while the $S_{cc'}^{\text{fl}}$ part gives rise to fluctuations in cross-sections

$$\langle \sigma_{cc'} \rangle = \sigma_{cc'}^{\text{DI}} + \langle \sigma_{cc'}^{\text{fl}} \rangle \tag{8.90}$$

where

$$\langle \sigma_{cc'}^{\text{fl}} \rangle = \frac{\pi}{k_c^2} \frac{2\pi}{D} \frac{\langle |\delta a|^2 \rangle}{\Gamma}. \tag{8.91}$$

The energy correlation function defined by

$$F(\varepsilon) = \langle (\sigma(E+\varepsilon) - \langle \sigma \rangle)(\sigma(E) - \langle \sigma \rangle) \rangle \tag{8.91}$$

is an important quantity. Using eqn (8.87) and the relation between the cross-section and the scattering matrix and using certain assumptions about δa_i it can be shown that

$$F_{cc'}(\varepsilon) = \frac{1}{1 + \left(\dfrac{\varepsilon}{\Gamma}\right)^2} \langle \sigma_{cc'}^{\text{fl}} \rangle (2\sigma_{cc'}^{\text{DI}} + \langle \sigma_{cc'}^{\text{fl}} \rangle). \tag{8.94}$$

The important conclusion is that $F(\varepsilon)$ has a Lorentzian form.

As we remarked earlier, an important feature of the S-matrix theory of nuclear reactions which is based on the R-matrix description is that the statistical properties of its parameters are directly related to the real eigenvalues and eigenfunctions of a many-body Hamiltonian. A full discussion of the fluctuation of cross-sections based on such a scattering matrix is quite involved; here we describe the calculation of the correlation function $F(\varepsilon)$ for the purely elastic scattering case using the pole resonance form given by (8.3a). For the purely elastic scattering case

$$F(\varepsilon) = \left(\frac{\pi}{k^2}\right)^2 [(\langle S(E+\varepsilon)S^*(E) \rangle + \text{c.c.}) \\ - (\langle S(E+\varepsilon) \rangle \langle S^*(E) \rangle + \text{c.c.})] \tag{8.94}$$

where c.c. denotes the complex conjugate.

The energy integrals are carried out using the Lorentzian weighting function as was done earlier for the average reaction cross-section. This gives

$$\langle S(E+\varepsilon)S^*(E) \rangle + \text{c.c.} = -i \sum_{\mu=1}^{N} \frac{g_\mu^2}{E + \varepsilon + \dfrac{iI}{2} - Z_\mu} \prod_{\lambda=1}^{N} \frac{\varepsilon - Z_\mu + Z_\lambda}{\varepsilon - Z_\mu + Z_\lambda^*}$$

$$-i \sum_{\mu=1}^{N} \frac{g_\mu^2}{E + \dfrac{iI}{2} - Z_\mu} \frac{-\varepsilon - Z_\mu + Z_\lambda}{-\varepsilon - Z_\mu + Z_\lambda^*} + \text{c.c.} + 2, \tag{8.95}$$

$$\langle S(E+\varepsilon) \rangle = 1 - i \sum_{\mu=1}^{N} \frac{g_\mu^2}{E + \varepsilon + \dfrac{iI}{2} - Z_\mu} \tag{8.96}$$

where we have suppressed the phase factor outside S. Using (8.94) and some further approximations,

$$F(\varepsilon) = 2\left(\frac{\pi}{k^2}\right)^2 \frac{2\pi}{D} \left\langle \frac{\Gamma_\mu}{1 + \varepsilon^2/\Gamma_\mu^2} \right\rangle. \tag{8.97}$$

Thus we see that shape of $F(\varepsilon)$ is again a Lorentzian.

8.6 Some general remarks and current developments

We have shown how the S-matrix is connected with the R-matrix and how to study the distribution of its parameters. Many of the correlations of S-matrix parameters have been experimentally verified using the vast amount of data on compound nucleus reactions. The technique of calculating correlations is quite general and can be used to study more complicated correlations, such as the correlations between three or more resonances. Since the correlations enter into the study of average cross-sections, we showed how the S-matrix can be used to study average cross-sections. In all these discussions we used the real quantities E_λ and $\gamma_{\lambda c}$. Theories can also be developed for the scattering matrix using the complex boundary-value problem (Moldauer 1964). This has certain advantages. These theories give the pole resonance form of the scattering matrix directly, but to study the statistical properties of these new S-matrices we have again to connect their parameters with another matrix, the statistical properties of which can be easily found. We discuss the averages over the space of complex orthogonal matrices in Chapter 11.

More recently (Verbaarschot et al. 1985) the theory of compound nucleus cross-sections has been studied using Grassmann algebra and the resolvent. These studies have the advantage that no expansions in terms of T/D have to made; thus they can provide more exact results for average cross-sections. Further details will be given in Chapter 10.

8.7 References

Anderson, T. W. (1958). *An introduction to multivariate statistical analysis*, Chapter II. John Wiley & Sons, New York.

Ericson, T. (1963). *Annals of Physics* (New York and London) **23**, 390.

Lane, A. M. and Thomas, R. G. (1958). *Review of Modern Physics* **30**, 257.

Mahaux, C. and Weidenmüller, H. A. (1969). *Shell model approach to nuclear reactions*. North-Holland, Amsterdam.

Mehta, M. L. (1967). *Random matrices*. Academic Press, New York.

Mehta, M. L. and Gaudin, M. (1960). *Nuclear Physics* **18**, 420.

Moldauer, P. A. (1964). *Physical Review* **B135**, 642.

Porter, C. E. (1965). *Statistical theories of spectra: fluctuations*. Academic Press, New York.

Preston, M. A. (1962). *Physics of the nucleus.* Addison-Wesley, Reading, Massachusetts.
Sandhya Devi, K. R. and Ullah, N. (1972). *Journal of Mathematical Physics* **13,** 1546.
Ullah, N. (1967). *Journal of Mathematical Physics* **8,** 1895.
Ullah, N. and Warke, C. S. (1968). *Physical Review* **170,** 857.
Verbaarschot, J. J. M., Weidenmüller, H. A., and Zirnbauer, M. R. (1985). *Physics Reports* **129,** 367.
Wigner, E. P. (1957). In *Proceedings of the Canadian Mathematical Congress,* p. 174. University of Toronto Press, Toronto.

9

HIGHLY CONVERGENT EXPANSION OF THE RAYLEIGH–SCHRÖDINGER PERTURBATION SERIES

9.1 Introduction

Since the beginning of quantum mechanics it has been realized that there are very few physical problems for which exact eigenfunctions and eigenvalues can be found analytically. In most cases some kind of perturbation theory must be used to calculate the eigenfunctions and eignevalues. approximately. (Merzbacher 1961). For many physical problems it turns out that a straightforward application of the perturbation series results in a divergent expansion of the desired eigenvalue. Under such circumstances a different method of calculating the approximate eigenvalue or some other technique to sum the series is used. From time to time, therefore, we need to find highly convergent expansions of the perturbation series. Recently, two new techniques (Feranchuk and Komarov 1982; Ullah 1985) have been used to achieve this goal, and in this chapter will discuss them. As is well known the anharmonic oscillator solution is needed in many branches of physics and oridinary perturbation theory starts diverging as the strength of the anharmonic term increases (Hioe and Montroll 1975; Singh et al. 1978). In fact in some cases we like to know the asymptotic solution when the coupling parameter goes to infinity. As shown in the next two sections both techniques give this asymptotic solution with very good accuracy. In §9.2 we describe the method known as the operator method (Feranchuk and Komarov 1982) and in §9.3 we describe the other method which splits the perturbation (Ullah 1985) into an average and fluctuating part.

9.2 Operator method

The essential idea of the operator method (Feranchuk and Komarov 1982) is to write the Hamiltonian in the normal form using annihilation and creaction operators. The unperturbed Hamiltonian is then taken to be that part of the Hamiltonian which commutes with the number operator.

Let us consider the anharmonic oscillator, for which the Hamiltonian

$$H = \tfrac{1}{2}(p^2 + x^2) + \lambda x^4 \tag{9.1}$$

where p and x are operators obeying the usual commutation relation

$$[x, p] = i, \quad (\hbar = 1).$$

and λ is a constant. Now use the transformation

$$x = \frac{1}{(2\omega)^{\frac{1}{2}}}(a + a^+), \tag{9.2a}$$

$$p = i\left(\frac{\omega}{2}\right)^{\frac{1}{2}}(a^+ - a), \tag{9.2b}$$

where $[a^+, a] = 1$ and ω is a parameter, and write the Hamiltonian H in terms of a^+ and a in the normal form,

$$H = \frac{\omega}{4}[1 - (a^+)^2 + 2a^+a - a^2] + \frac{1}{4\omega}[1 + (a^+)^2 + 2a^+a + a^2]$$

$$+ \frac{\lambda}{4\omega^2}[3 + 6(a^+)^2 + 12a^+a + 6a^2 + (a^+)^4 + 4(a^+)^3a + 6(a^+)^2a^2$$

$$+ 4a^+a^3 + a^4]. \tag{9.3}$$

We now construct H_0 such that all terms in H_0 commute with a^+a, giving

$$H_0 = \tfrac{1}{4}\omega + \frac{1}{4\omega} + \frac{3\lambda}{4\omega^2} + \left(\tfrac{1}{2}\omega + \frac{1}{2\omega} + \frac{3\lambda}{\omega^2}\right)(a^+a) + \frac{3\lambda}{\omega^2}(a^+)^2a^2. \tag{9.4}$$

The energy spectrum of H_0 is

$$E_n^{(0)}(\omega) = \left(\tfrac{1}{2}\omega + \frac{1}{2\omega} + \frac{3\lambda}{\omega^2}\right)(n + \tfrac{1}{2}) + \frac{3\lambda}{\omega^2}(n^2 - n - \tfrac{1}{2}). \tag{9.5}$$

The unknown parameter ω is determined for each n by minimizing $E_n^{(0)}(\omega)$. This gives

$$\omega_n^3 - \omega_n - 6\lambda \frac{(n^2 + n + \tfrac{1}{2})}{n + \tfrac{1}{2}} = 0. \tag{9.6}$$

With this solution for the parameter ω for each n, the zero-order energy

$$E_n^{(0)} = \tfrac{1}{4}\left(3\omega_n + \frac{1}{\omega_n}\right)(n + \tfrac{1}{2}). \tag{9.7}$$

Denoting the remaining part of the Hamiltonian H as a perturbation V where

$$V = H - H_0, \tag{9.8}$$

we see that the first-order correction is zero. The second- and higher-order corrections can be calculated in the usual way. Thus, the expression

9.3 METHOD BASED ON THE LINEARIZATION TECHNIQUE

for the second-order correction $E_0^{(2)}$ is

$$E_0^{(2)} = -3\lambda^2/4\omega_0^2(2\omega_0 + 21\lambda). \tag{9.9}$$

Explicit expressions for $E_0^{(n)}$ for the first few values of n are given in Appendix A. The ground-state energy up to fourth order for $\lambda = 1$ is

$$E_0^{(0)} + E_0^{(2)} + E_0^{(4)} = 0.803747.$$

The exact value as given by Hioe and Montroll (1975) is 0.803771. Thus we see that the approximate eigenvalue is very close to its exact value.

One could also calculate the asymptotic form of the ground-state energy when $\lambda \to \infty$. The leading term up to second-order is

$$E_0^{(0)} + E_0^{(2)} \underset{\lambda \to \infty}{=} 0.66953 \lambda^{\frac{1}{3}},$$

which again is very close to the exact value (Hioe and Montroll 1975) of E_0,

$$E_0 \underset{\lambda \to \infty}{=} 0.667986 \lambda^{\frac{1}{3}}.$$

9.3 Method based on the linearization technique

We again consider the same Hamiltonian H for the anharmonic oscillator given by (9.1). The essential idea of the linearization techniques (Ullah 1985) is to approximate the term x^4 as $\langle x^2 \rangle x^2$, where $\langle x^2 \rangle$ is the expectation value of x^2 with respect to some vacuum state. The linearization technique is used quite often in the random phase approximation (Rowe 1970) which is used to calculate the vibrational states of nuclei. We, therefore, write the Hamiltonian H given by (9.1) as

$$H = H_0 + H_1 \tag{9.10}$$

where

$$H_0 = -\frac{1}{2} \frac{\partial^2}{\partial x^2} + \frac{1}{2}(1 + 2\beta)x^2 \tag{9.11}$$

and

$$H_1 = \lambda x^4 - \beta x^2. \tag{9.12}$$

β is an unknown parameter to be determined. The eigenfunctions of H_0 can be immediately written as

$$\psi_n = N_n H_n(b^{\frac{1}{2}} x) \exp(-\tfrac{1}{2} b x^2), \tag{9.13a}$$

$$b = (1 + 2\beta)^{\frac{1}{2}}, \tag{9.13b}$$

$$N_n = \left[\frac{b^{\frac{1}{2}}}{\pi^{\frac{1}{2}} 2^n n!} \right]^{\frac{1}{2}}. \tag{9.13c}$$

The next problem is to determine the unknown parameter β. From our knowledge of many-body physics we know that the largest configuration interaction (Nesbet 1955) effects arise from the lowest configurations. We, therefore, determine β by postulating that the matrix element of H_1 between ψ_0 and ψ_2 must vanish. This gives a cubic equation

$$\beta^2(1+2\beta) - 9\lambda^2 = 0. \tag{9.14}$$

The contributions to ground-state energy in various orders of perturbation can now be easily calculated, giving

$$E_0^{(0)} = \tfrac{1}{2}(1+2\beta)^{\frac{1}{2}}, \tag{9.15a}$$

$$E_0^{(1)} = -3\lambda/4(1+2\beta), \tag{9.15b}$$

$$E_0^{(2)} = -\tfrac{3}{8}\lambda^2/(1+2\beta)^{\frac{5}{2}}. \tag{9.15c}$$

We again calculate the ground-state energy when the coupling constant $\lambda = 1$. For this case eqn (9.14) gives $\beta = \tfrac{3}{2}$. The ground-state energy up to the second order of perturbation using relations (9.15) is 0.801 which is quite close to the exact value 0.803771 quoted in § 9.2.

For the asymptotic case we find that (9.14) gives

$$\beta \underset{\lambda \to \infty}{=} (\tfrac{9}{2})^{\frac{1}{3}}\lambda^{\frac{2}{3}}. \tag{9.16}$$

Using (9.15), we find that the asymptotic value of ground state energy up to the second order of perturbation is

$$E_0 = \tfrac{35}{48}(\tfrac{3}{4})^{\frac{1}{3}}\lambda^{\frac{1}{3}}$$

$$\underset{\lambda \to \infty}{=} 0.66\lambda^{\frac{1}{3}},$$

which again is quite close to its exact value of $0.67\lambda^{\frac{1}{3}}$.

We shall now comment on the method presented above. We see that an alternative way to look at the method is to re-write the given perturbation by taking out a part which is used to redefine the zero-order Hamiltonian. Thus, in cases where linearization is not possible, we could still modify the perturbation to get a rapid convergence. We could also introduce more than one parameter and use other parameters to further minimize the configuration interaction effects. As we have just seen, the method enables us to find a good approximation to the given eigenvalue using the smallest number of low-order configurations.

9.4 Concluding remarks

The techniques which we described here are extremely useful because they enable us to calculate the eigenvalues using low orders of perturbation theory in cases where ordinary perturbation theory starts diverging.

At present the methods are applied to single-particle Hamiltonians. It is expected that in future such methods will also be extended to the many-body problem.

9.5 References

Feranchuk, I. D. and Komarov, L. I. (1982). *Physics Letters* **88A,** 211.
Hioe, F. T. and Montroll, E. W. (1975). *Journal of Mathematical Physics* **16,** 1945.
Merzbacher, E. (1961). *Quantum mechanics.* John Wiley & Sons, New York.
Nesbet, R. K. (1955). *Proceedings of the Royal Society, London* **A230,** 312.
Rowe, D. J. (1970). *Nuclear collective motion.* Methuen, London.
Singh, V., Biswas, S. N., and Datta, K. (1978). *Physical Review* **D18,** 1901.
Ullah, N. (1985). *Pramana* **24,** 27.

10
APPLICATION OF GRASSMANN INTEGRATION IN MATRIX ENSEMBLE THEORY

10.1 Introduction

As mentioned in Chapter 9, the matrix ensemble theory was introduced by Wigner to study the average properties of the compound nucleus resonances and their widths. This theory is finding new applications, e.g. in the study of conductivity of electrons in a random potential (Ramond 1981; Efetov 1982) and the effect of random potential on well-behaved spectra (Verbaarschot et al. 1984a). In these studies new techniques have been developed to study the ensemble averages of the Hamiltonian matrix elements. As we showed in Chapter 9, the joint eigenvalue distribution of the eigenvalues for the Gaussian orthogonal ensemble (GOE) case includes a factor involving the absolute value of the differences of eigenvalues. The absolute value sign makes further integrations difficult. To overcome this the method of integrating over alternate variables was developed by Mehta (1967). An alternative way is to start from the resolvent and find its ensemble average. The simplest way to evaluate these averages is to make use of Grassman integration (Ramond 1981; Efetov 1982). In § 10.2 we shall give a brief introduction to Grassmann integration and then make use of this algebra to obtain an expression for the average resolvent in § 10.3. Some other results concerning the Fourier transform of a single-eigenvalue distribution and the average compound nucleus cross-section will also be given in § 10.3.

10.2 Grassmann integration

The Grassman algebra or the algebra of the anticommuting variables (Ramond 1981; Efetov 1982; Balian and Zinn-Justin 1976) is currently used in several areas of theoretical physics. We will describe the important rules of this algebra and illustrate them using several simple examples. Some of these results will be used in the next section.

Consider the anticommuting variables a^* and a obeying the rules (Balian and Zinn-Justin 1976)

$$a^*a + aa^* = 0, \tag{10.1a}$$

$$a^{*2} = 0, \tag{10.1b}$$

$$a^2 = 0. \tag{10.1c}$$

10.2 GRASSMANN INTEGRATION

They are the generators of Grassmann algebra. A general function $f(a^*, a)$ can be written

$$f(a^*, a) = f_0 + \tilde{f}_1 a + f_1 a^* + f_2 a a^* \tag{10.2}$$

where f_0, $\tilde{f}_1$, f_1, and f_2 are complex numbers.

The basic rules of integrations as given by Berezin are

$$\int da^* = 0, \tag{10.3a}$$

$$\int da = 0, \tag{10.3b}$$

$$\int a^* \, da^* = 1, \tag{10.3c}$$

$$\int a \, da = 1. \tag{10.3d}$$

da^* and da anticommute with each other as well as with a^* and a. We shall now use these rules to work out the integrals over Grassmann variables for a number of cases.

First consider the integral

$$\int aa^* \exp(-a^*a) \, da^* \, da. \tag{10.4}$$

If we expand the exponential, then from (10.1) we get

$$\exp(-a^*a) = 1 - a^*a.$$

The integral (10.4) then becomes

$$\int aa^*(1 - a^*a) \, da \, da.$$

Again the second term vanishes because of the property of anticommuting variables given by (10.1), while the application of rules (10.3) gives value

$$\int aa^* \exp(-a^*a) \, da^* \, da = 1.$$

As a second example, consider the integral

$$I_2(M) = \exp - \left[(a_1 a_2) \begin{pmatrix} 0 & m_{12} \\ -m_{12} & 0 \end{pmatrix} \begin{pmatrix} a_1 \\ a_2 \end{pmatrix} \right] da_1 \, da_2.$$

$\begin{pmatrix} a_1 \\ a_2 \end{pmatrix}$ is a two-dimensional vector and the matrix

$$M = \begin{pmatrix} 0 & m_{12} \\ -m_{12} & 0 \end{pmatrix}$$

is a two-dimensional antisymmetric matrix.

It is easy to see that $I_2(M)$ can be written

$$I_2(M) = \int \exp(-2m_{12}a_1a_2)\, da_1\, da_2,$$

which gives, using the rules of integration,

$$I_2(M) = 2m_{12} = 2(\det M)^{\frac{1}{2}}$$

where det M is the determinant of matrix M.

This result can be generalized for N dimensions to

$$I_N(M) = 2^{\frac{N}{2}}(\det M)^{\frac{1}{2}}$$

provided N is even; otherwise zero.

We must point out here that in writing the integration rules (10.3) some authors write $\int da\, a = 1$ instead of $\int a\, da = 1$. Also in the theory of compound nucleus cross-sections, for convenience $\int a\, da$ is taken to be $(2\pi)^{-\frac{1}{2}}$. These conventions do not affect the final results provided one follows them consistently.

As a final example, consider a slightly more difficult integral (Ramond 1981; Efetov 1982)

$$I_2(M; \chi) = \int \exp\left[-(a_1 a_2)\begin{pmatrix} 0 & m_{12} \\ -m_{12} & 0 \end{pmatrix}\begin{pmatrix} a_1 \\ a_2 \end{pmatrix} + \sum_{i=1}^{2} \chi_i a_i\right] da_1\, da_2$$

where χ_i anticommute among themselves as well as with a_i.

Using the property of anticommutativity,

$$I_2 = \int \exp(-2m_{12}a_1a_2 + \chi_1 a_1 + \chi_2 a_2)\, da_1\, da_2.$$

On expanding the exponential and using integration rules,

$$I_2 = 2(m_{12} + \tfrac{1}{2}\chi_1\chi_2).$$

Introducing the inverse matrix

$$M^{-1} = \begin{pmatrix} 0 & \dfrac{-1}{m_{12}} \\ \dfrac{1}{m_{12}} & 0 \end{pmatrix},$$

10.3 PROBABILITY DENSITY OF A SINGLE EIGENVALUE

we can write I_2 as

$$I_2(M;\chi) = \left[\exp-\tfrac{1}{4}(\chi_1\chi_2)M^{-1}\begin{pmatrix}\chi_1\\\chi_2\end{pmatrix}\right]I_2(M).$$

The most important result which can be proved using the above rules of Grassmann integration and which is very useful in the theory of matrix ensembles is that the determinant of a matrix A

$$\det A = \int \exp(-a^*Aa)\,da^*\,da \qquad (10.5)$$

where $a_1, \ldots, a_n, a_1^*, \ldots, a_n^*$ are $2n$ anticommuting variables.

10.3 Probability density of a single eigenvalue

For physical systems the relevant matrix ensemble is the Gaussian Orthogonal Ensemble (GOE). We consider a real symmetric $N \times N$ Hamiltonian matrix, the elements of which have the distribution

$$P(H) = K\exp(-\operatorname{tr} H^2) \qquad (10.6)$$

where K is the normalization constant and tr denotes the trace of the matrix.

To derive the single-eigenvalue probability density function we first calculate the average resolvent. The resolvent $G(Z)$, used extensively in many-body physics, is

$$G(Z) = \frac{1}{N}\sum_{i=1}^{N}\frac{1}{Z-E_i}, \qquad (10.7a)$$

$$G(Z) = \frac{1}{N}\operatorname{tr}(Z-H)^{-1} \qquad (10.7b)$$

where E_i are the eigenvalues.

To calculate the average value of the resolvent we write it in terms of determinants. Using (10.7a) or (10.7b), $G(Z)$ can be written

$$G(Z) = \frac{1}{N}\frac{\frac{\partial}{\partial\xi}\det(\xi-H)}{\det(Z-H)}\bigg|_{\xi=Z}. \qquad (10.8)$$

Since the distribution of H given by eqn (10.6) is Gaussian, we represent the determinants appearing in eqn (10.8) as

$$(\det M)^{-\frac{1}{2}} = (\pi)^{-N/2}\int_{-\infty}^{\infty}\exp\left(-\sum_{i,j}X_iX_jM_{ij}\right)\prod_{i}^{N}dX_i \qquad (10.9)$$

where M is a real symmetric matrix, and

$$\det M = \exp\left(-\sum_{i,j} a_i^* M_{ij} a_j\right) \prod_i^N da_i^* \, da_i \tag{10.10}$$

where a_i^* and a_i are the anticommuting Grassmann variables.

The ensemble average $g(Z)$ of the resolvent $G(Z)$ given by eqn (10.8) can now be written using eqns (10.6), (10.9), and (10.10). Carrying out integrations over H_{ij} variables,

$$g(Z) = \frac{1}{N} \frac{1}{i^N} \left(\frac{\partial}{\partial \xi}\right) \frac{1}{\pi^N} \int \prod_k dx_k \, dy_k \prod_k da_k^* \, da_k$$

$$\times \exp\left[-\tfrac{1}{4} \sum (x_k^2 + y_k^2)^2 + iZ \sum (x_k^2 + y_k^2)\right.$$

$$\left. + \sum_{k<j} (x_k x_j + y_k y_j)^2 \right] \exp\left[-\xi \sum a_k^* a_k - \frac{i}{2} \sum_k (x_k^2 + y_k^2)\right.$$

$$\times a_k^* a_k + \tfrac{1}{2} \sum_{k<j} a_k^* a_k a_j^* a_j + i \sum_{k<j} (x_k x_j + y_k y_j)$$

$$\left. \times (a_k^* a_j + a_j^* a_k)\right]_{\xi=Z}. \tag{10.11}$$

We now specialize to the case of two dimensions. For this case, integrating over Grassmann variables,

$$g(Z) = \left(-\frac{1}{2\pi^2}\right)\left(2Z + \frac{1}{2}\frac{\partial}{\partial Z}\right) \int dx_1 \, dx_2 \, dy_1 \, dy_2$$

$$\times \exp\{-\tfrac{1}{4}[(x_1^2 + y_1^2)^2 + (x_2^2 + y_2^2)^2] + iZ(x_1^2 + y_1^2$$

$$+ x_2^2 + y_2^2) + (x_1 x_2 + y_1 y_2)^2\}. \tag{10.12}$$

Using polar coordinates and after some further simplifications, the final expression for $g(Z)$ is

$$g(Z) = \frac{2}{i} \int_0^\infty \exp(-t^2 + 2itZ) M\left(-\tfrac{1}{2}, \tfrac{1}{2}, \tfrac{t^2}{2}\right) dt \tag{10.13}$$

where $M(a, b, x)$ is the confluent hypergeometric function (Abramowitz and Stegun 1965). It is now a simple matter to find the probability density function of the single eigenvalue, by noting that (10.7b) represents Stieltje's integral of moments. Thus the probability density function

$$P(E) = \frac{2}{\pi} \int_0^\infty dt \, \exp(-t^2) \cos(2tE) M\left(-\tfrac{1}{2}, \tfrac{1}{2}, \tfrac{t^2}{2}\right) \tag{10.14}$$

which, on integrating over t, gives (Gradshteyn and Ryshik 1965)

$$P(E) = (2\pi)^{-\frac{1}{2}} \exp(-E^2) M(-\tfrac{1}{2}, \tfrac{1}{2}, -E^2). \tag{10.15}$$

10.3 PROBABILITY DENSITY OF A SINGLE EIGENVALUE

If desired the single-eigenvalue probability density function given by (10.5) can be expressed in terms of the error function by using the relation between the confluent hypergeometric function and the error function (Abramowitz and Stegun 1965).

It is interesting to note here that new approximations to $P(E)$ can be obtained using expressions of the form of (10.13). Thus, if we retain the first two terms in the expansion of the confluent hypergeometric function in eqn (10.13) we get the approximate single-eigenvalue probability density function

$$P_{\text{ap}}(E) = \tfrac{3}{4} \pi^{\frac{1}{2}} \exp(-x^2)(1 + \tfrac{2}{3} x^2). \tag{10.16}$$

This is a very good approximation to the exact expression given by (10.15) as can be seen by calculating the value of

$$\frac{\int P(E)P_{\text{ap}}(E)\,dE}{([\int P^2(E)\,dE][\int P_{\text{ap}}^2(E)\,dE]^{\frac{1}{2}}},$$

which turns out to be 0.97.

Before we proceed further, we mention that Dyson had classified the matrix ensembles into three classes. (1) the Gaussian Orthogonal Ensemble (GOE) which has a real symmetric Hamiltonian; (2) the Gaussian Unitary Ensemble (GUE) in which the Hamiltonian matrix is Hermitian; and (3) the Gaussian symplectic Ensemble (GSE) in which the Hamiltonian matrix is a quarternion. We would now like to show that a general expression can be obtained for the Fourier transform of the single-eigenvalue probability density function for the three ensembles. Rather than using the resolvent, we derive this transform using the method of moments.

Consider a real symmetric Hamiltonian H. It can be easily shown that

$$\det(1 - \lambda H) = \exp\left[-\sum_{k=1}^{\infty} \frac{\lambda^k}{k} \operatorname{tr} H^k \right] \tag{10.17}$$

where λ is a parameter. This determinantal identity can be used to calculate the ensemble averages of the traces of powers of H. Taking the log of (10.17) and writing the average using the distribution (10.6),

$$-\sum_{k=1}^{\infty} \frac{\lambda^k}{k} \langle \operatorname{tr} H^k \rangle = \left[\int \exp -(H_{11}^2 + H_{22}^2 + 2H_{12}^2) \prod_{i \leq j} dH_{ij} \right]^{-1}$$

$$\times \left[\int \ln[1 - \lambda(H_{11} + H_{22}) + \lambda^2(H_{11}H_{22} - H_{12}^2)] \right.$$

$$\left. \times \exp -(H_{11}^2 + H_{22}^2 + 2H_{12}^2) \prod_{i \leq j} dH_{ij} \right]. \tag{10.18}$$

Using first the transformation $H_{11} + H_{22} = u$, $H_{11} - H_{22} = v$ and then putting $v = \rho \cos \theta$, and $2H_{12} = \rho \sin \theta$, after some simplification,

$$\langle \operatorname{tr} H^k \rangle = 2^{-k} \left[\int_{\rho=0}^{\infty} \int_{u=-\infty}^{\infty} [\exp -\tfrac{1}{2}(u^2 + \rho^2)] \rho \, d\rho \, du \right]^{-1}$$

$$\left[\int_{\rho=0}^{\infty} \int_{u=-\infty}^{\infty} \exp -\tfrac{1}{2}(u^2 + \rho^2) \right] [(u+\rho)^k + (u-\rho)^k] \rho \, d\rho \, du]. \quad (10.19)$$

It is obvious from eqn (10.3) that, if k is odd, the ensemble-averaged trace vanishes. For $k = 2m$, we can carry out the integrations in (10.19) and obtain the expression

$$\langle \operatorname{tr} H^{2m} \rangle = \frac{\Gamma(m + \tfrac{1}{2})}{\pi^{\frac{1}{2}} 2^{m-1}} F(-m, 1; \tfrac{1}{2}; -1) \quad (10.20)$$

where F denotes the hypergeometric function (Abramowitz and Stegun 1965).

For the Gaussian unitary ensemble (GUE), the joint distribution of the Hamiltonian matrix elements is

$$P_{\text{unitary}}(H) = K \exp - \operatorname{tr}(HH^\dagger). \quad (10.21)$$

Following the same procedure we find for GUE

$$\langle \operatorname{tr} H^{2m} \rangle = \frac{\Gamma(m + \tfrac{1}{2})}{\pi^{\frac{1}{2}} 2^{m-1}} F(-m, \tfrac{3}{2}; \tfrac{1}{2}; -1), \quad (10.22)$$

and a similar expression for GSE.

Since we now know all the moments of $(\operatorname{tr} H^k)$ we can calculate the characteristic function $\phi(t)$ given by

$$\phi(t) = \sum_{r=0}^{\infty} \frac{(it)^r}{r!} M_r \quad (10.23)$$

where M_r denote the moments. First consider GUE. Using eqns (10.22) and (10.23) we find for GUE

$$\phi(t) = \sum_{m=0}^{\infty} \frac{1}{m!} \left(-\frac{t^2}{8} \right)^m F(-m, \tfrac{3}{2}; \tfrac{1}{2}; -1). \quad (10.24)$$

Using the formula which gives the sum (Hansen 1975) over hypergeometric functions,

$$\phi(t) = \exp\left(-\frac{t^2}{4}\right) M\left(-1, \tfrac{1}{2} \frac{t^2}{8}\right) \quad (10.25)$$

where M is the confluent hypergeometric function. Expression (10.25) gives the Fourier transform for the two-dimensional GUE.

10.3 PROBABILITY DENSITY OF A SINGLE EIGENVALUE

We can now write the Fourier transform for all three Gaussian ensembles by introducing parameter β which has values 1, 2, and 4 for GOE, GUE, and GSE, respectively. The diagonal Hamiltonian matrix elements are now taken to have the variance $\sigma^2 = 1/\beta$, while the off-diagonal has variance $\frac{1}{2}$ of the diagonal. The Fourier transform $\phi_\beta(t)$ for all three cases is given by

$$\phi_\beta(t) = \exp\left(-\frac{t^2}{2\beta}\right) M\left(-\frac{\beta}{2}, \frac{1}{2}, \frac{t^2}{4\beta}\right). \tag{10.26}$$

We now consider the case when the dimension N of the Hamiltonian matrix is very large. Special techniques (Verbaarschot et al. 1984b) employing saddle points are used for large N, some of which will be mentioned when we discuss the average value of the scattering matrix later.

For large N we find that the ensemble average of the resolvent for GOE is

$$g(Z) = \left[\frac{2}{N}(Z - (Z^2 - N)^{\frac{1}{2}})\right]. \tag{10.27}$$

It is interesting to note here that an equation for $g(Z)$ can be derived by expanding eqn (10.7b) and resumming the dominant terms of expansion, giving

$$g = \frac{4}{4Z - Ng}, \tag{10.28}$$

which was first given by Pastur (1972).

As in the derivation of (10.14) from (10.13), we can now easily derive Wigner's semicircular distribution from eqn (10.27) giving

$$P(E) = \frac{2}{N\pi}(N - E^2)^{\frac{1}{2}}, \quad |E| < N^{\frac{1}{2}}. \tag{10.29}$$

A very interesting application of Grassmann integration is that of Verbaarschot, Weidenmüller, and Zirnbauer (1984b) who calculated the averages of the scattering matrix without relying on the expansion in terms of average width to average spacing. We shall now give some of the details of this application and also give their final expression for the average of the product of two S-matrix elements at two different energies.

Let us consider a set of N compound-nuclear levels $|\mu\rangle$ which are coupled by a Hamiltonian $H_{\mu\nu}$. It is assumed that $H_{\mu\nu}$ is real symmetric and therefore belongs to GOE. Further, the ensemble averages of H are assumed to be

$$\langle H_{\mu\nu}\rangle = 0, \quad \langle H_{\mu\nu} H_{\mu'\nu'}\rangle = \frac{\lambda^2}{N}(\delta_{\mu\mu'}\delta_{\nu\nu'} + \delta_{\mu\nu'}\delta_{\nu\mu'}). \tag{10.30}$$

The scattering matrix S is given by

$$S_{ab}(E) = \delta_{ab} - 2i\pi \sum_{\mu,\nu} W_{\mu a}(D^{-1})_{\mu\nu} W_{\nu b} \qquad (10.31)$$

where a, b denote the channels, E denotes the energy, and $W_{\mu a}$ describes the coupling between level μ and channel a. It is important to note that $W_{\mu a}$ unlike the R-matrix amplitudes are taken to be fixed. The inverse propagator $D_{\mu\nu}$ is given by

$$D_{\mu\nu} = E\delta_{\mu\nu} - H_{\mu\nu} + i\pi \sum_{a} W_{\mu a} W_{\nu a}. \qquad (10.32)$$

Just as in chapter 8, the phase shift factors are suppressed in writing expression (10.30). Further, direct coupling between the channels is neglected. This enables us to write

$$\langle S_{ab}(E) \rangle = \delta_{ab} \langle S_{aa}(E) \rangle \qquad (10.33)$$

where the bracket sign $\langle \ \rangle$ denotes the average over GOE. These ensemble averages are calculated at fixed energy.

The goal is to express the ensemble average of the product of two S-matrix elements at two different energies E_1 and E_2 in terms of the transmission coefficients T_c which are related to optical model phase shifts and are given by

$$T_c = 1 - |\langle S_{cc} \rangle|^2. \qquad (10.34)$$

This goal is achieved by first writing a generating function for the propagator and then carrying out the ensemble averages over GOE.

In dealing with a problem which involves both ordinary variables denoted by $S_i (i = 1, \ldots, 2L)$ and Grassman variables χ_i, χ_i^* ($i = 1, \ldots, L$) it is convenient to introduce vectors ϕ of the form

$$\phi = \begin{pmatrix} S_1 \\ \vdots \\ S_{2L} \\ \chi_1 \\ \vdots \\ \chi_L \\ \chi_1^* \\ \vdots \\ \chi_L^* \end{pmatrix},$$

which are called graded vectors.

10.3 PROBABILITY DENSITY OF A SINGLE EIGENVALUE

The scalar product of such a vector with its Hermitian adjoint $\phi^\dagger$ is

$$(\phi^\dagger \cdot \phi) = \sum_{\mu=1}^{2L} S_\mu^2 + 2 \sum_{\mu=1}^{L} \chi_\mu^* \chi_\mu$$

Similarly one introduces $4L \times 4L$ dimensional matrices called graded matrices F of the form

$$F = \begin{pmatrix} a & \sigma \\ \rho & b \end{pmatrix}$$

where a and b are $2L \times 2L$ matrices having ordinary variables as their matrix elements while σ and ρ are $2L \times 2L$ matrices having anticommuting variables as their matrix elements.

In terms of these, the generating function $Z(E, J)$ is given (Verbaarschot et al. 1985) by

$$Z(E, J) = \int d(\phi) \exp[\mathscr{L}(\phi, J)] \tag{10.35}$$

where $d[\phi]$ is the volume element in the space of $2N$ real ordinary variables and $2N$ anticommuting Grassmann variables and ϕ is the graded vector, the first $2N$ components of which are ordinary variables and the last $2N$ anticommuting variables. The Lagrangian $\mathscr{L}(\phi, J)$ denotes the scalar product

$$\mathscr{L} = \tfrac{1}{2} i(\phi^\dagger D' \phi) \tag{10.36}$$

where D' is a $4N \times 4N$ graded matrix, consisting of blocks of matrices $D(J^C)$, $D(J^C)$, $D(J^A)$, and $D(J^A)$, which are defined

$$D_{\mu\nu}(J^C) = D_{\mu\nu} - J^C_{\mu\nu}, \tag{10.37a}$$

$$D_{\mu\nu}(J^A) = D_{\mu\nu} + J^A_{\mu\nu} \tag{10.37b}$$

where J^A and J^C are symmetric. The superscripts C and A denote commuting and anticommuting variables respectively.

Using the integration rules for the Grassmann algebra it can be shown that

$$Z(E, J) = \det(D(J^A))[\det(D(J^C))]^{-1}. \tag{10.38}$$

The importance of the generating function arises from the fact that we can write the propagator, say as

$$(Z - \delta_{\mu\nu})(D^{-1})_{\mu\nu} = \frac{\partial}{\partial J^C_{\mu\nu}} Z(E, J) \Big|_{J^A = J^C = 0}. \tag{10.39}$$

We can now calculate the ensemble averages of matrix elements of S

using eqns (10.30)–(10.32), (10.35)–(10.37), and (10.39). After making use of the Hubbard–Stratonivitch transformation and saddle point integration and after considerable algebraic manipulation, we can show that the ensemble average of the product of two matrix elements of S can be written as a threefold integral. The final result is

$$\langle S_{ab}((E(1)S_{cd}^*(E(2)))\rangle - \langle S_{ab}\rangle\langle S_{cd}^*\rangle$$
$$= \tfrac{1}{8}\int_0^\infty d\lambda_1 \int_0^\infty d\lambda_2 \int_0^1 d\lambda$$
$$\times \frac{(1-\lambda)\lambda\,|\lambda_1 - \lambda_2|}{((1+\lambda_1)\lambda_1(1+\lambda_2)\lambda_2)^{\frac{1}{2}}(\lambda+\lambda_1)^2(\lambda+\lambda_2)^2}$$
$$\times \exp\{i\pi(E_1 - E_2^*)d^{-1}(\lambda_1 + \lambda_2 + 2\lambda)\}$$
$$\times \prod_e \frac{1 - T_e\lambda}{(1 + T_e\lambda_1)^{\frac{1}{2}}(1 + T_e\lambda_2)^{\frac{1}{2}}}$$
$$\times \left\{\delta_{ab}\delta_{cd}\langle S_{aa}\rangle\langle S_{cc}^*\rangle T_a T_c\left(\frac{\lambda_1}{1+T_a\lambda_1} + \frac{\lambda_2}{1+T_a\lambda_2} + \frac{2\lambda}{1-T_a\lambda}\right)\right.$$
$$\times \left(\frac{\lambda_1}{1+T_c\lambda_1} + \frac{\lambda_2}{1+T_c\lambda_2} + \frac{2\lambda}{1-T_c\lambda}\right) + (\delta_{ac}\delta_{bd} + \delta_{ad}\delta_{bc})T_a T_b$$
$$\times \left(\frac{\lambda_1(1+\lambda_1)}{(1+T_a\lambda_1)(1+T_b\lambda_1)} + \frac{\lambda_2(1+\lambda_2)}{(1+T_a\lambda_2)(1+T_b\lambda_2)}\right.$$
$$\left.\left. + \frac{2\lambda(1-\lambda)}{(1-T_a\lambda)(1-T_b\lambda)}\right)\right\} \tag{10.40}$$

where d is the average spacing.

In general eqn (10.40) cannot be integrated in a closed form and can only be evaluated numerically. However, Verbaarschot (1986) showed that the integrations in (10.40) can be carried out for certain special cases, e.g. when the number of channels are large and the transmission coefficients for each channel are small. For this case we can make the approximations

$$\frac{1}{1+T_c\lambda_1} \to 1, \quad \frac{1}{1+T_c\lambda_2} \to 1, \quad \frac{1}{1-T_c\lambda} \to 1$$

and

$$\prod_c \frac{(1-T_c\lambda)}{(1+T_c\lambda_1)(1+T_c\lambda_2)} \to \exp - t(\lambda + \tfrac{1}{2}\lambda_1 + \tfrac{1}{2}\lambda_2), \quad t = \sum T_c.$$

After making a number of approximations and using certain results from the GOE two-point function $G_{\mu\nu}^{(E_1)} G_{\mu'\nu'}^*(E_2)$, eqn (10.40) for the

10.3 PROBABILITY DENSITY OF A SINGLE EIGENVALUE

case of small transmission coefficients can be written as

$$\langle S_{ab}(E_1)S_{cd}^*(E_2)\rangle - \langle S_{ab}\rangle\langle S_{cd}^*\rangle$$
$$= \delta_{ab}\delta_{cd}T_aT_c\left[\frac{1}{8}\int_0^2 dS\, S(2-\log(1+S))\right.$$
$$\times \exp\left.-S\left(\frac{t}{2}+ir\right)\right] + \frac{1}{8}\int_2^\infty dS(4-S\ln\frac{S+1}{S-1})$$
$$\times \exp\left[-S\left(\frac{t}{2}+ir\right)\right] + (\delta_{ad}\delta_{bc}+\delta_{ac}\delta_{bd})\frac{T_aT_b}{(t+2ir)} \quad (10.41)$$

where $r = \pi(E_2^* - E_1)/d$ and the variable t has been introduced for analytic continuation.

We can now calculate, as an example the elastic enhancement factor W defined by

$$W = \frac{\langle |S_{aa}|^2\rangle}{\langle |S_{ab}|^2\rangle}\bigg|_{a\neq b}, \quad r = 0. \quad (10.42)$$

Using (10.41), we find that

$$W = 2 + \frac{t}{4}\int_0^2 dS\, S(2-\ln(S+1))\exp\left(-\frac{ts}{2}\right)$$
$$+ \frac{t}{4}\int_2^\infty ds\left(4-S\ln\frac{S+1}{S-1}\right)\exp\left(-\frac{ts}{2}\right). \quad (10.43)$$

Expanding $4 - S\ln[(S+1)/(S-1)]$ in a Taylor series expansion and keeping the first term, we find that in the limit $t\to 0$, W has the value 3. This value of the elastic enhancement factor for small transmission coefficients agrees with the value obtained earlier by Moldauer using a different method.

The other case which can be discussed using eqn (10.40) is that of overlapping resonances for which one expands the integrals in inverse powers of the sum of transmission coefficients. This is also shown (Verbaarschot 1986) to agree with the earlier results obtained in the study of average compound nucleus cross-sections.

Numerical integration is used to calculate the correlation function

$$C(r) = \frac{\langle S_{ab}(E-\varepsilon)S_{ab}^*(E)\rangle}{\langle |S_{ab}(E)|^2\rangle} \quad (10.44)$$

as a function of the energy difference ε measured in units of the local average spacing d. The curves thus obtained for $C(r)$ can be fitted by Lorentzians; however, the asymptotic Lorentzian form

$$C_L(r) = \frac{t^2}{t^2 + 4r^2}$$

is not even completely reached for a case with 20 equivalent channels and $t = \sum T_c = 20.0$.

We have so far discussed the averages of the product of two S-matrix elements at two different energies. For a complete study of the fluctuation of cross-sections, we need the ensemble averages of the product of four S-matrix elements which are not available at present but in principle can be worked out using Grassmann integration and the generating functions $Z(E, J)$.

We also mention here that the resolvent given by eqn (10.7b) is also called the one-point function. As was shown in this section it is very useful in calculating the single-eigenvalue probability density function. For the calculation of the correlation of two eigenvalues, we introduce the two-point correlation function, an example of which was mentioned when we discussed the elastic enhancement factor. The two-point correlation function can be written as

$$g(Z, \xi) = \left\langle \frac{1}{N} \text{tr}(Z - H)^{-1} \frac{1}{N} \text{tr}(\xi - H)^{-1} \right\rangle. \tag{10.45}$$

Expressing $g(Z, \xi)$ in terms of the eigenvalues, we can write (10.45) as

$$g(Z, \xi) = \frac{1}{N} \left\langle \frac{1}{(Z - E_1)(\xi - E_1)} \right\rangle + \frac{N-1}{N} \left\langle \frac{1}{(Z - E_1)(\xi - E_2)} \right\rangle. \tag{10.46}$$

We see from eqn (10.46) that the first part of $g(Z, \xi)$ is already known from the one-point function, while the second part gives information about the correlations between two eigenvalues. The two-point correlation function was used (Verbaarschot and Zirnbauer 1984) to find Dyson's two-level cluster function.

We have already discussed how to use Grassmann variables to express a determinant in exponential form. The exponential form of the determinant makes it easier to carry through the ensemble averages since the distribution of the Hamiltonian matrix elements is given as Gaussian. In the last part of this chapter we would like to consider whether other simpler representations could serve this purpose. We know from the theory of angular momentum (Edmonds 1957; Brink and Satchler 1962) that the step-up and step-down operators when operating on angular momentum eigenfunctions annihilate them when the absolute value of the projection quantum number exceeds the j-quantum number. This fact can be used to construct a representation of determinants in terms of step-up operators. We show the results for small dimensions.

Consider a 2×2 real symmetric matrix A and consider the matrix element M of the step-up operators l_+ in the basis of spherical harmonics $Y_{2\mu}$

10.3 PROBABILITY DENSITY OF A SINGLE EIGENVALUE

$$M = \tfrac{1}{24} \langle Y_{22}(2)Y_{22}(1)| \exp(A_{11}l_+^2(1) + A_{22}l_+^2(2) + i2^{\frac{1}{2}}A_{12}l_+(1)l_+(2) |Y_{20}(2)Y_{20}(1)\rangle. \tag{10.47}$$

It we now expand the exponential, we find that because of the annihilation property of l_+, the highest power in the expansion of the exponential will be 2. The factor $1/24$, has been inserted for normalization purposes; it could have been added later. Expanding the exponential up to the second power and writing the matrix elements of l_+^2 between spherical harmonics (Edmonds 1957; Brink and Satchler 1962), we find that

$$M = \det A. \tag{10.48}$$

Thus eqn (10.47) provides a representation of the determinant of a 2×2 real symmetric matrix.

We next consider a 3×3 real symmetric matrix A. In this case we can show that, if we write

$$M = (24)^{-3/2} \operatorname{Re} \langle Y_{22}(1)Y_{22}(2)Y_{22}(3)|$$
$$\exp\left[\sum_{i=1}^{3} A_{ii}l_+^2(i) + C \sum_{i<j} A_{ij}l_+(i)l_+(j)\right]$$
$$|Y_{20}(1)Y_{20}(2)Y_{20}(3)\rangle \tag{10.49}$$

where Re stands for the real part and C is complex,

$$C = a + ib,$$

then

$$M = \det A, \tag{10.50}$$

provided

$$b^2 = a^2 + 2, \tag{10.51a}$$

and a is the root of

$$a^3 + 3a - 1 = 0. \tag{10.51b}$$

Thus, for small dimensions, $N = 2, 3$, we can represent the determinant of a real symmetric matrix in terms of step-up operators.

We now show an application of the representation given by (10.48) in calculating the ensemble averages of the traces of powers of Hamiltonian matrix. We first write the identity (10.17) as

$$\sum_{k=1}^{\infty} \langle \operatorname{tr} H^k \rangle \lambda^{k-1} = \langle \det(1 - \lambda H) \frac{\partial}{\partial \lambda} \frac{1}{\det(1 - \lambda H)} \rangle. \tag{10.52}$$

Assuming that the distribution of H is given by (10.6), the ensemble average indicated by (10.52) can be easily taken using eqns (10.9) and (10.48). We can, therefore, write

$$\sum_{k=1}^{\infty} \langle \mathrm{tr}\, H^k \rangle \lambda^{k-1}$$

$$= \frac{1}{2\pi^2} \left(-\frac{\partial}{\partial \alpha}\right) \langle Y_{22}^{(1)} Y_{22}^{(2)} | \left[\exp(l_+^2(1) + l_+^2(2) - \frac{\alpha^2}{4} l_+^2(1) l_+^2(2) \right]$$

$$\times \int dx\, dy\, du\, dv \exp\{-(x^2 + y^2 + u^2 + v^2)\}$$

$$\times \exp\{\tfrac{1}{4}[\lambda^2(u^2+x^2)^2 + (v^2+y^2)^2 + 2(uv+xy)^2]$$
$$- 2\alpha\lambda[(u^2+x^2)l_+^2(1) + (v^2+y^2)l_+^2(2)]$$
$$- 2(2)^{\frac{1}{2}} i\alpha\lambda[(uv+xy)l_+(1)l_+(2)]\} |Y_{20}(2) Y_{20}(1)\rangle_{\alpha=\lambda}. \qquad (10.53)$$

In writing eqn (10.53) all terms in the exponential having powers of l_+ greater than two have been omitted because they cannot connect Y_{22} and Y_{20}. Integrating over u, x, v, and y by expressing $\exp\tfrac{1}{4}\lambda^2(u^2+x^2)^2$ as $\exp\tfrac{1}{2}\lambda(u^2+x^2)$, etc., writing the matrix elements of l_+ operators, and differentiating with respect to α, after a few simplifying steps, we can write

$$\sum_{k=1}^{\infty} \langle \mathrm{tr}\, H^k \rangle \lambda^{k-1} = \frac{\lambda}{2\pi^{\frac{1}{2}}} \int_{-\infty}^{\infty} du \int_0^{\infty} \rho\, d\rho [1 - 2^{\frac{1}{2}}\lambda u + \tfrac{1}{2}\lambda^2 u^2 - \tfrac{1}{2}\lambda^2 \rho^2]^{-2}$$
$$\times [(6-\lambda^2) - 4(2)^{\frac{1}{2}}\lambda u - x^2\rho^2 + \lambda^2 u^2] \exp-(u^2+\rho^2). \qquad (10.54)$$

As earlier, if k is odd, $\langle \mathrm{tr}\, H^k \rangle$ vanishes, while if $k = 2m$, we find by integrating over u, ρ and after a few simplifying steps that

$$\langle \mathrm{tr}\, H^{2m} \rangle = \frac{1}{\pi^{\frac{1}{2}}} \frac{\Gamma(m+\tfrac{1}{2})}{2^{m-1}} [(2mF(-m+1, 1; \tfrac{3}{2}; -1) + 1]. \qquad (10.55)$$

Using Gauss's relations for contiguous functions it can be easily shown that

$$\langle \mathrm{tr}\, H^{2m} \rangle = \frac{1}{\pi^{\frac{1}{2}}} \frac{\Gamma(m+\tfrac{1}{2})}{2^{m-1}} F(-m, 1; \tfrac{1}{2}; -1), \qquad (10.56)$$

which is precisely the result derived earlier and given by eqn (10.20).

10.4 References

Abramowitz, M. and Stegun, A. (1965). *Handbook of mathematical functions.* Dover, New York.

10.4 REFERENCES

Balian, R. and Zinn-Justin, J. (Eds.) (1976). In *Methods in field theory, Les Houches École d'Étude Physique Théorique*, 1975, Session 28, p. 1. North-Holland, Amsterdam.
Brink, D. M. and Satchler, G. R. (1962). *Angular momentum*. Oxford University Press, Oxford.
Edmonds, A. R. (1957). *Angular momentum in quantum mechanics*. Princeton University Press, Princeton, New Jersey.
Efetov, K. B. (1982). *Soviet Physics—JETP* **55,** 514.
Gradshteyn, I. S. and Ryzhik, I. M. (1965). In *Tables of integrals, series and products* (ed. A. Jeffrey). Academic Press, New York and London.
Hansen, E. R. (1975). *A table of series and products*. Prentice-Hall, Englewood Cliffs, New Jersey.
Mehta, M. L. (1967). *Random matrices*. Academic Press, New York.
Pastur, L. A. (1972). *Teor. Mat. Fiz.* **10,** 102.
Ramond, P. (1981). *Field theory*, p. 214. Benjamin/Cummings, Reading, Massachusetts.
Verbaarschot, J. J. M. (1986). *Annals of Physics* (New York) **168,** 368.
Verbaarschot, J. J. M., Weidenmüller, H. a., and Zirnbauer, M. (1984a). *Annals of Physics* (New York) **153,** 367.
Verbaarschot, J. J. M., Weidenmüller, H. A., and Zirnbauer, M. R. (1984b). *Physics Letters* **B149,** 263.
Verbaarschot, J. J. M., Weidenmüller, H. A., and Zirnbauer, M. R. (1985). *Physics Reports* **129,** 367.
Verbaarschot, J. J. M. and Zirnbauer, M. R. (1984). *Annals of Physics* (New York) **158,** 78.

11
AVERAGES OVER THE SPACE OF COMPLEX ORTHOGONAL MATRICES

11.1 Introduction

In studying the fluctuation of cross-sections, the most suitable form of the scattering matrix for calculating its energy averages is the pole resonance form which has resonance parameters which are almost energy independent. We showed in Chapter 9 how the scattering matrix is related to the R-matrix which has all real parameters which are related to the many-body Hamiltonian. To express the scattering matrix S using R-matrix theory with real boundary conditions, it is by no means simple to write it in the pole resonance form and connect the parameters of the pole resonance form with those of the real R-matrix. It can be done for the purely elastic scattering case. As the number of channels increase, it becomes more and more difficult to write the pole resonance form. Because of this Moldauer (1946a) introduced what he called a statistical collision matrix which is a good approximation to the actual collision matrix over a small energy interval. In order to write the pole resonance form in a simple way, the complex boundary conditions are used. In order to study the statistical properties of the parameters of the statistical collision matrix we must study the averages over the space of the complex orthogonal matrix. In § 11.2 we briefly describe the statistical collision matrix and study the averages of the parameters of this matrix using the volume element in the space of complex orthogonal matrices in § 11.3. The average cross-sections based on the statistical collision matrix are described in § 11.4.

11.2 Statistical collision matrix

The statistical collision matrix $U^s(E, E_0)$ is written (Moldauer 1964a) as

$$U^s(E, E_0) = U^0(E_0) - i \sum_\mu \frac{g_\mu \times g_\mu}{E - \varepsilon_\mu + \frac{i}{2}\Gamma_\mu} \qquad (11.1)$$

where E_0 is some specified total energy and $U^0(E_0)$ is the part that gives rise to the direct reactions. $g_\mu \times g_\mu$ is the matrix formed from the complex amplitudes which are defined in terms of the complex amplitudes θ_μ of

11.2 STATISTICAL COLLISION MATRIX

the R-matrix. ε_μ and $-\frac{1}{2}\Gamma_\mu$ are the real and imaginary parts of the complex eigenvalues W_μ of the eigenvalue equation

$$HX_\mu = W_\mu X_\mu \tag{11.2}$$

where H is the many-body Hamiltonian and X_μ are the internal wave functions. The eigenvalue eqn (11.2) is solved by specifying boundary conditions which are complex numbers B_c. Defining $\psi_{\mu c}$ and $\theta_{\mu c}$ by the relations

$$\theta_{\mu c} = \left(\frac{\hbar^2}{2M_c a_c}\right)^{\frac{1}{2}} \int \phi_c^* X_\mu \, dS \tag{11.3a}$$

where dS is an element of channel surface and

$$\psi_{\mu c} = \left(\frac{\hbar^2}{2M_c a_c}\right)^{\frac{1}{2}} \int \phi_c^* \nabla_n (r_c X_\mu) \, dS \tag{11.3b}$$

where as earlier, M_c is the reduced mass in channel c, a_c the channel radius, and ∇_n denotes the normal gradient at the surface. The complex numbers B_c are related to $\psi_{\mu c}$, and $\theta_{\mu c}$ by the relation

$$\frac{\psi_{\mu c}}{\theta_{\mu c}} = B_c. \tag{11.4}$$

Assuming H to be invariant under rotations and under time reversal, eqns (11.2) and (11.4) give

$$H\tilde{X}_\mu = W_\mu^* \tilde{X}_\mu, \tag{11.5}$$

$$\frac{\tilde{\psi}_{\mu c}}{\tilde{\theta}_{\mu c}} = B_c^* \tag{11.6}$$

where $\sim$ denotes time-reversed states which can be written

$$\tilde{X}_\mu(J, M) = (-1)^{J-M} K X_\mu(J, J-M) \tag{11.7}$$

where K is the time reversal operator (Wigner 1959) and J and M denote the total angular momentum quantum number and its projection.

From the above relations it is easy to establish that X_μ and $\tilde{X}_\nu$ ($\mu \neq \nu$) are orthogonal. The wave functions X_μ are assumed to be normalized according to

$$\int_{\text{interior}} \tilde{X}_\mu^* X_\nu \, d\tau = \delta_{\mu\nu}. \tag{11.8}$$

Denoting by N_μ the integral

$$N_\mu = \int_{\text{interior}} |X_\mu|^2 \, d\tau \tag{11.9}$$

where interior denotes the internal region, it can be shown that

$$\Gamma_\mu = \sum_c \Gamma_{\mu c}, \tag{11.10}$$

$$\Gamma_{\mu c} = 2 N_\mu^{-1} \operatorname{Im} B_c |\theta_{\mu c}|^2, \tag{11.11}$$

where Im denotes the imaginary part.

The R-matrix for the complex boundary problem is defined

$$R_{cc'} = \sum_\mu \frac{\theta_{\mu c} \theta_{\mu c'}}{\varepsilon_\mu - E - \tfrac{1}{2} i \Gamma_\mu}. \tag{11.12}$$

The complex amplitudes $g_{\mu c}$ are given by

$$g_{\mu c} = \Omega_c (2 P_c)^{\tfrac{1}{2}} \theta_{\mu c} \tag{11.13}$$

where P_c is the penetration factor for channel c. Denoting by I_c and O_c the incoming and outgoing solutions of the radial equation

$$\left[\frac{d^2}{dr_c^2} - \frac{l(l+1)}{r_c^2} - \frac{2 M_c}{\hbar^2} (V - E) \right] u = 0, \tag{11.14}$$

Ω_c and P_c are defined as

$$\Omega_c = \left(\frac{I_c}{O_c} \right)^{\tfrac{1}{2}} \tag{11.15a}$$

$$L_c = \frac{k_c a_c O_c'}{O_c} = S_c + i P_c. \tag{11.15b}$$

11.3 Averages of the parameters of the statistical collision matrix

In this section we would like to study the statistical properties of the complex amplitudes $\theta_{\mu c}$ of the R-matrix given by eqn (11.12). Using the orthogonality property expressed by the relation (11.8), we shall show that the statistical properties of $\theta_{\mu c}$ will be connected with a complex orthogonal matrix. As in Chapter 9 where we had real amplitudes and therefore needed averages over the space of real orthogonal matrices we shall now need averages over the space of complex orthogonal matrices.

Our starting point is again the expansion of the internal wave function in terms of a convenient orthonormal basis set (Ullah 1967).

$$X_\mu = \sum_\nu a_{\nu\mu} \Phi_\nu \tag{11.16}$$

where $a_{\nu\mu}$ $(1 \leq \nu \leq N)$ are the components of the eigenvector belonging to eigenvalue W_μ. It has been shown (Porter 1965) that, if the Hamiltonian is invariant under rotations as well as under time reversal,

11.3 THE STATISTICAL COLLISION MATRIX

the basis set Φ_ν can be chosen so that

$$K\Phi_\nu = \Phi_\nu \tag{11.17}$$

where K is the time reversal operator (Wigner 1959).

We next show that the matrix formed from the coefficients $a_{\nu\mu}$ is a complex orthogonal matrix. To show this we substitute the expansion of X_μ given in eqn (11.16) in the orthogonality relation (11.8). This together with (11.17) gives

$$\sum_\alpha \tilde{a}_{\mu\alpha} a_{\alpha\nu} = \delta_{\mu\nu} \tag{11.18}$$

which shows that the matrix formed from the elements $a_{\mu\nu}$ is a complex orthogonal matrix.

The statistical distribution of any quantity which is a function of $a_{\mu\nu}$ can now be worked out provided we know the volume element in the space of complex orthogonal matrices. Because of certain difficulties we first consider the simplest case of two dimensions, which will be very useful when we take up the general case of N dimensions.

It is easy to see that a 2×2 complex orthogonal matrix a can be parametrized as

$$a = \begin{pmatrix} \cos\omega & \sin\omega \\ -\sin\omega & \cos\omega \end{pmatrix} \tag{11.19}$$

where $\omega = \omega_1 + i\omega_2$.

We now calculate the line element dS^2 given by

$$dS^2 = \text{tr } da\, da\dagger, \tag{11.20}$$

which using the parametrized form (11.19) can be written as

$$dS^2 = [\exp(2\omega_2) + \exp(-2\omega_2)](d\omega_1^2 + d\omega_2^2). \tag{11.21}$$

Thus, the volume element $\dot{a}$ can be written as

$$\dot{a} = \exp(2\omega_2) + \exp(-2\omega_2)\, d\omega_1\, d\omega_2 \tag{11.22a}$$

where

$$-\pi \leq \omega_1 \leq \pi; \quad -\infty \leq \omega_2 \leq \infty. \tag{11.22b}$$

From relations (11.14) and (11.15) we see that the total volume of the complex orthogonal space is not bounded and that, therefore, the normalization integral will diverge. This was the difficulty mentioned earlier which does not arise in the case of real orthogonal matrices. What it means is that we cannot take the probability density $P(a)$ proportional to $\dot{a}$ throughout the range indicated by (11.22b). The solution to this difficulty is to introduce a weighting factor which will ensure the convergence of the normalization integral. The two most suitable forms which we consider here are: (a) the unit step function in the range $-q \leq \omega_2 \leq q$; and (b) the Gaussian weighting factor $\exp(-p\omega_2^2)$. The values of the parameter p or q will be obtained by comparing one of the calculated quantities with its known value.

We now calculate the moments of the complex amplitudes $\theta_{\mu c}$ given by eqn (11.3a). Using the expansion (11.16), $\theta_{\mu c}$ can be written

$$\theta_{\mu c} = \sum_v a_{v\mu} J_{vc} \qquad (11.23a)$$

where

$$J_{vc} = \left(\frac{\hbar^2}{2M_c a_c}\right)^{\frac{1}{2}} \int_{\text{surface}} \phi_c^* \Phi_v \, dS. \qquad (11.23b)$$

From the fact that

$$\tilde{\theta}_{\mu c}^* = \theta_{\mu c}$$

it is easy to see that J_{vc} is real.

Using eqns (11.19), (11.22), and (11.23) we find for case (b)

$$\langle |\theta_{\mu c}|^{2n} \rangle_\mu = (\tfrac{1}{2} J_c)^n \exp(-p^{-1}) \sum_{s=0}^{n} \left\{ \left[\binom{n}{s}\right]^2 \exp[(n - 2s + 1)^2 p^{-1}] \right\}, \qquad (11.24)$$

where $\langle \ \rangle_\mu$ denotes the ensemble average, $J_c = \tfrac{1}{2}(J_{1c}^2 + J_{2c}^2)$, and $\binom{n}{s}$ is the binomial coefficient.

For case (a), if $n = 2m$, we have

$$\langle |\theta_{\mu c}|^{4m} \rangle_\mu = (\tfrac{1}{2} J_c)^{2m} [\exp(2q) - \exp(-2q)]^{-1}$$

$$\times \sum_{s=0}^{2m} \left\{ \left[\binom{2m}{s}\right]^2 (2m - 2s + 1)^{-1} \right.$$

$$\left. \times [\exp 2q(2m - 2s + 1) - \exp[-2q(2m - 2s + 1)]] \right\} \qquad (11.25a)$$

and, if $n = 2m + 1$,

$$\langle |\theta_{\mu c}|^{4m+2} \rangle_\mu = \tfrac{1}{2}(\tfrac{1}{2} J_c)^{2m+1} [\exp(2q) - \exp(-2q)]^{-1}$$

$$\times \left(\sum_{s=0}^{2m+1} \left\{ \left[\binom{2m+1}{s}\right]^2 (m - s + 1)^{-1} \right. \right.$$

$$\left. \times [\exp[4q(m - s + 1)] - \exp[-4q(m - s + 1)]] \right\}$$

$$\left. + 8\left[\binom{2m+1}{m+1}\right]^2 q \right) \qquad (11.25b)$$

11.3 THE STATISTICAL COLLISION MATRIX

where a prime on the summation over s indicates that the term $s = m + 1$ has to be excluded.

It is interesting to see that, if we take the limit $p \to \infty$ in (11.24) or $q \to 0$ in (11.25), we recover the result of the real boundary condition

$$\langle |\theta_{\mu c}|^{2n} \rangle_\mu = (\tfrac{1}{2} J_c)^n \binom{2n}{n}. \tag{11.26}$$

In the study of average cross-sections (Moldauer 1964a) which will be given later, we introduce two parameters B_c and A_c defined by the relations

$$B_c = \left[\frac{|\langle \theta_{\mu c}^2 \rangle_\mu|}{\langle |\theta_{\mu c}^2|\rangle_\mu} \right]^2, \tag{11.27a}$$

$$A_c = \frac{\langle |\theta_{\mu c}|^4 \rangle_\mu}{[\langle |\theta_{\mu c}|^2 \rangle_\mu]^2}. \tag{11.27b}$$

Using the above ensemble averages, these two parameters can be expressed in terms of a single parameter p or q. We find for case (a)

$$B_c = 16[\exp(2q) - \exp(-2q)]^2 [\exp(4q) \\ - \exp(-4q) + 8q]^{-2} \tag{11.28a}$$

$$A_c = \tfrac{4}{3}[\exp(2q) - \exp(-2q)]^2 \\ \times [\exp(4q) - \exp(-4q) + 8q]^{-2} \\ \times [\exp(4q) + \exp(-4q) + 16] \tag{11.28b}$$

and for case (b)

$$B_c = 4 \exp(2p^{-1})[\exp(4p^{-1}) + 1]^{-2}, \tag{11.29a}$$

$$A_c = \exp(2p^{-1})[\exp(4p^{-1}) + 1]^{-2}[\exp(8p^{-1}) + 5]. \tag{11.29b}$$

Again, if we take the limiting process $p \to \infty$ or $q \to 0$, we find that $A_c \to 1.5$ which is the result we get for the real boundary conditions. We would now like to generalize the results for an arbitrary number of dimensions N of the complex orthogonal matrix. As with real orthogonal matrices it is difficult to write in general an explicit parametrized form of the real orthogonal matrix and to calculate the volume element starting from the square of the line element. Instead we introduce δ-functions to take into account the conditions imposed by normalization and the orthogonality of rows or columns. The same procedure can be used for complex orthogonal matrices.

First consider a single column vector of the N-dimensional random complex orthogonal matrix. Since we are considering a single column vector, we shall suppress the column index β from the component $a_{\alpha\beta}$.

The real and imaginary parts of a_α are denoted by a^{Re} and a^{Im}, respectively. Using the δ-function to take into account the normalization condition

$$\sum_{\alpha=1}^{N} a_\alpha^2 = 1, \tag{11.30}$$

the ensemble average of any quantity Q which is a function of a^{Re} and a^{Im} can be written

$$\langle Q \rangle = K^{-1} \int Q(\{a_\alpha^{Re}, a_\alpha^{Im}\}) \delta\left(\sum_{\alpha=1}^{N} (a_\alpha^{Re})^2 - \sum_{\alpha=1}^{N} (a_\alpha^{Im})^2 - 1\right)$$
$$\times \delta\left(\sum_{\alpha=1}^{N} a_\alpha^{Re} a_\alpha^{Im}\right) \prod_{\alpha=1}^{N} da_\alpha^{Re} da_\alpha^{Im}, \tag{11.31}$$

where $\langle \ \rangle$ denotes the ensemble average and K is the normalization integral, the same integral as given by (11.31) with $Q = 1$.

The earlier discussion of the two-dimensional case tells us that the integral in eqn (11.31) will diverge and we have to introduce a weighting function for convergence. The simplifying assumption which we now make is that the weigthing function ρ_N is of the form $\rho_N(\sum_{\alpha=1}^{N} (a_\alpha^{Im})^2)$. Introducing this weighting function in eqn (11.31) and making a simple transformation, we can rewrite it as

$$\langle Q \rangle = K^{-1} \int Q(\{(1+\lambda)^{\frac{1}{2}} u_\alpha, \lambda^{\frac{1}{2}} v_\alpha\})$$
$$\times [\lambda(1+\lambda)]^{\frac{1}{2}(N-3)} \rho_N(\lambda)$$
$$\times \delta\left(\sum u_\alpha^2 - 1\right) \delta\left(\sum v_\alpha^2 - 1\right) \delta\left(\sum u_\alpha v_\alpha\right) d\lambda \prod_\alpha du_\alpha \, dv_\alpha. \tag{11.32}$$

We would now like to show that the above formulation enables us to obtain relations between the parameters of the collision matrix without any detailed knowledge of the weighting function $\rho_N(\lambda)$. For this purpose we consider the complex amplitudes $\theta_{\mu c}$ given by eqn (11.23a), the complex pole residues $g_{\mu c}$ given by (11.13), the normalization constant

$$N_\mu = \sum_\nu |a_{\nu\mu}|^2, \tag{11.33}$$

and the parameters B_c and A_c given by (11.27).

Using these relations it can be shown that

$$B_c = [\langle N_\mu \rangle]^{-2}. \tag{11.34}$$

Further, the quantity $\Gamma_{\mu c}$ defined by eqn (11.11) can be written as

$$\Gamma_{\mu c} = \frac{|g_{\mu c}|^2}{N_\mu}. \tag{11.35}$$

11.3 THE STATISTICAL COLLISION MATRIX

Table 11.1 Comparison of the values of B_c and $\langle \Gamma_{\mu c} \rangle$ with their numerically calculated values

No. of channels	$\langle N_\mu \rangle$	$\langle |g_{\mu c}|^2 \rangle$	B_c Numerical	B_c Present	$\langle \Gamma_{\mu c} \rangle$ Numerical	$\langle \Gamma_{\mu c} \rangle$ Present
20	1.18	0.144	0.53	0.72	0.097	0.122
100	1.52	0.108	0.39	0.43	0.064	0.071
300	1.69	0.081	0.33	0.35	0.047	0.048

We also introduce the quantity $\Theta_{\mu c}$ defined by the relation

$$\Theta_{\mu c} = \frac{2\pi}{D} N_\mu |g_{\mu c}|^2 \tag{11.36}$$

where D is the mean spacing.

Again using the ensemble average of Q and relations defining $g_{\mu c}$ and N_μ, we can show that

$$\langle \Gamma_{\mu c} \rangle = \frac{\langle |g_{\mu c}|^2 \rangle}{\langle N_\mu \rangle}. \tag{11.37}$$

We would now like to see how relations (11.34) and (11.37) check with the numerical calculations carried out by Moldauer (1964b). In these calculations the parameters of the real boundary value problem (Lane and Thomas 1958) are used and, by numerically diagonalizing a complex symmetric level matrix, the averages of the parameters of the statistical collision matrix are obtained. In Table 11.1 we give the values of B_c and $\langle \Gamma_{\mu c} \rangle$ assuming the values of $\langle N_\mu \rangle$ and $\langle |g_{\mu c}|^2 \rangle$ and compare them with the ones obtained using numerical calculations. We see that the agreement between the two set of values is quite good, the agreement being better as the number of channels increases.

Other relations of this type are

$$A_c = N(N+2)^{-1}[1 + 2\langle N_\mu^2 \rangle][\langle N_\mu \rangle]^{-2} \tag{11.38a}$$

and

$$\langle \Theta_{\mu c} \rangle = \frac{2\pi}{D} \langle \Gamma_{\mu c} \rangle \langle N_\mu^2 \rangle. \tag{11.38b}$$

We next calculate the normalized mean-square deviation defined by the relation

$$S(X_\mu) = \frac{\langle X_\mu^2 \rangle - \langle X_\mu \rangle^2}{\langle X_\mu \rangle^2}. \tag{11.39}$$

We find that the normalized mean-square deviation of $|g_{\mu c}|^2$ is

$$S(|g_{\mu c}|^2) = (N+2)^{-1}[N-2+2NS(N_\mu)+N(\langle N_\mu\rangle)^{-2}]. \quad (11.40)$$

Thus the normalized mean-square deviation of $|g_{\mu c}|^2$ can be expressed in terms of the normalized mean-square deviation of N_μ. If we assume the values of $S(N_\mu)$ to be given, we can calculate the value of $S(|g_{\mu c}|^2)$. Again it is found that the values calculated using eqn (11.40) agree very well with values obtained by numerical calculations.

It is interesting to note that for the real boundary value problem eqn (11.38a) becomes

$$A_c = \frac{3N}{N+2},$$

which for large values of N becomes 3 in agreement with the value of A_c obtained using the Porter–Thomas (P–T) distribution of the partial width (Porter 1965).

We would next like to work out the distribution of the absolute square of the amplitude $\theta_{\mu c}$. This derivation is useful because it shows how to work out the distribution of the various parameters of the statistical collision matrix using integrals of the form given by eqn (11.32).

Using (11.23a), we can write

$$|\theta_{\mu c}|^2 = \left(\sum_\alpha a_\alpha^{\text{Re}} J_{\alpha c}\right)^2 + \left(\sum_\alpha a_\alpha^{\text{Im}} J_{\alpha c}\right)^2. \quad (11.41)$$

For the derivation of the distribution of $|\theta_{\mu c}|^2$, it will be convenient to define a quantity

$$X = \frac{|\theta_{\mu c}|^2}{J_c} \quad (11.42a)$$

where

$$J_c = \frac{1}{N}\sum_\alpha J_{\alpha c}^2. \quad (11.42b)$$

Using eqns (11.31), (11.32), (11.41), and (11.42) we can write the probability density function of X as

$$P_N(X) = k^{-1}\int \delta[(1+\lambda)(\sum u_\alpha J_{\alpha c})^2 + \lambda(\sum v_\alpha J_{\alpha c})^2 - XJ_c]$$

$$\times \delta(\sum u_\alpha^2 - 1)\delta(\sum v_\alpha^2 - 1)\delta(\sum u_\alpha v_\alpha)[\lambda(1+\lambda)]^{\frac{1}{2}(N-3)}$$

$$\times \rho_N(\lambda)\,d\lambda \prod_\alpha du_\alpha\,dv_\alpha \quad (11.43)$$

where k^{-1} is the appropriate normalization constant.

11.3 THE STATISTICAL COLLISION MATRIX

Making a real orthogonal transformation on the variables u_α and v_α,

$$u'_\beta = \sum_\alpha u_\alpha C_{\alpha\beta}, \tag{11.44a}$$

$$v'_\beta = \sum_\alpha v_\alpha C_{\alpha\beta} \tag{11.44b}$$

where C is orthogonal matrix and choosing

$$C_{\alpha 1} = \frac{J_{\alpha c}}{\left(\sum_\alpha J_{\alpha c}^2\right)^{\frac{1}{2}}}, \tag{11.45}$$

we can rewrite (11.44) as

$$P_N(X) = k' \int \delta\left[(1+\lambda)u_1'^2 + \lambda v_1'^2 - \frac{X}{N}\right]$$
$$\times \delta(\sum u_\alpha'^2 - 1)\delta(\sum v_\alpha'^2 - 1)\delta(\sum u_\alpha' v_\alpha')$$
$$\times [\lambda(1+\lambda)]^{\frac{1}{2}(N-3)} \rho_N(\lambda) \, d\lambda \prod_\alpha du_\alpha' \, dv_\alpha' \tag{11.46}$$

where k' is the new normalization constant. In writing (11.46) we used the fact that, under an orthogonal transformation, the volume element and the scalar products remain invariant.

Integrating over all u'_α and v'_α except u'_1 and v'_1 and calling the latter u, v we can write eqn (11.46) as

$$P_N(X) = k \int \delta\left[(1+\lambda)u^2 + \lambda v^2 - \frac{X}{N}\right][1 - u^2 - v^2]^{\frac{1}{2}(N-4)}$$
$$\times [\lambda(1+\lambda)]^{\frac{1}{2}(N-3)} \rho_N(\lambda) \, d\lambda \, du \, dv \tag{11.47}$$

where k is the new normalization integral and the u, v integration is inside the circle $(u^2 + v^2) \leq 1$.

Equation (11.47) is exact. Further integration over the variables u and v is difficult to carry out. We now make the approximation of large N which is the situation in practice. For large N we can approximately replace the factor $[1 - u^2 - v^2]^{\frac{1}{2}(N-4)}$ with $\exp(-\frac{1}{2}N(u^2 + v^2))$ and take the limits of integration over u, v from $-\infty$ to ∞. Using the Fourier transform of the δ-function and performing integrations over u, v,

$$P_N(X) = \left[2\int d\lambda \, \rho_N(\lambda)[\lambda(1+\lambda)]^{\frac{1}{2}(N-3)}\right]^{-1}$$
$$\times \int d\lambda \, \rho_N(\lambda)[\lambda(1+\lambda)]^{\frac{1}{2}(N-4)}$$
$$\times \exp\left[-\frac{X}{4} \cdot \frac{2\lambda+1}{\lambda(1+\lambda)}\right] I_0\left(\frac{X}{4\lambda(1+\lambda)}\right) \tag{11.48}$$

where I_0 is the modified Bessel function (Whittaker and Watson 1962) of the first kind.

To proceed further we need the form of the weighting function which is most simply obtained by writing an expression for the probability density function of N_μ and comparing it with the one obtained numerically.

Using eqns (11.32) and (11.33) and after a few algebraic steps,

$$P(N_\mu) = \left[\int_1^\infty (N_\mu^2 - 1)^{\frac{1}{2}(N-3)} \rho_N[\tfrac{1}{2}(n_\mu - 1)]\, dN_\mu\right]^{-1}$$
$$\times (N_\mu^2 - 1)^{\frac{1}{2}(N-3)} \rho_N[\tfrac{1}{2}(N_\mu - 1)], \qquad 1 \leq N_\mu \leq \infty. \quad (11.49)$$

This will fit the numerical curve for $P(N_\mu)$ if $\rho_N(\lambda)$ is chosen to be

$$\rho_N(\lambda) = [\lambda(1 + \lambda)]^{-\frac{1}{2}(N-5)} \exp(-a\lambda) \quad (11.50)$$

where a is a parameter.

With this choice of $\rho_N(\lambda)$, the density function

$$P(N_\mu) = \tfrac{1}{8}a^3(a + 2)^{-1}(N_\mu^2 - 1)\exp[-\tfrac{1}{2}a(N_\mu - 1)]. \quad (11.51)$$

Putting (11.50) in (11.48) gives

$$P_N(X) = \tfrac{1}{2}a^3(a+2)^{-1}\int_0^\infty d\lambda\, [\lambda(1+\lambda)]^{\frac{1}{2}}$$
$$\times \exp\left[-\left(a\lambda + \frac{X}{4}\frac{2\lambda + 1}{\lambda(1+\lambda)}\right)\right] I_0\left(\frac{X}{4\lambda(1+\lambda)}\right). \quad (11.52)$$

For large values of a and X not too small, we can expand I_0 for large argument and approximately evaluate the integral. This gives the probability density function of the absolute square of the complex amplitude $\theta_{\mu c}$ when X is not too small. The form so obtained closely resembles the P–T distribution of the real R-matrix theory.

The value of the parameter a can be obtained, e.g. by comparing the value of $\langle N_\mu \rangle$ calculated using eqn (11.51) with its value obtained using numerical calculation. We can then calculate the normalized mean-square deviations $S(N_\mu)$, $S(\Gamma_{\mu c})$, and $S(\Theta_{\mu c})$. They are given in Table 11.2 along with their values obtained by using numerical calculation. As can be seen the agreement between the two sets of values is fairly good.

As mentioned in Chapters 8 and 9, in the study of the average value and fluctuation of cross-sections the various correlations of the parameters of the scattering matrix must be known. We now briefly describe some of the correlations of the parameters of the statistical collision matrix. The calculation of various ensemble averages is similar to that given for the probability density function of the quantity X and therefore

11.3 THE STATISTICAL COLLISION MATRIX

Table 11.2 Comparison of the normalized mean-square deviations $S(N_\mu)$, $S(\Gamma_{\mu c})$, and $S(\Theta_{\mu c})$ assuming $\langle N_\mu \rangle$ to be given

No. of channels	$\langle N_\mu \rangle$	a	Numerical calculation			Present calculation		
			$S(N_\mu)$	$S(\Gamma_{\mu c})$	$S(\Theta_{\mu c})$	$S(N_\mu)$	$S(\Gamma_{\mu c})$	$S(\Theta_{\mu c})$
20	1.18	23.1	0.01	1.60	1.81	0.01	1.67	1.82
100	1.52	8.4	0.006	1.38	1.88	0.057	1.40	1.98
300	1.69	6.5	0.07	1.31	1.76	0.08	1.34	2.14

we shall omit the details of the calculation when we discuss the correlations. We first consider the channel–channel correlation of $|\theta_{\mu c}|^2$. Denoting this correlation coefficient by $\rho^{(c,c')}_{|\theta|^2}$, we write

$$\rho^{(c,c')}_{|\theta|^2} = \frac{\langle |\theta_{\mu c}|^2 |\theta_{\mu c'}|^2 \rangle - \langle |\theta_{\mu c}|^2 \rangle \langle |\theta_{\mu c'}|^2 \rangle}{[(\langle |\theta_{\mu c}|^4 \rangle - \langle |\theta_{\mu c}|^2 \rangle^2)(\langle |\theta_{\mu c'}|^4 \rangle - \langle |\theta_{\mu c'}|^2 \rangle^2)]^{\frac{1}{2}}} \quad (11.53)$$

The ensemble averages needed in expression (11.53) can be shown to be

$$\langle |\theta_{\mu c}|^2 |\theta_{\mu c'}|^2 \rangle = J_c J_{c'} [\langle (2\lambda+1)^2 \rangle + 2p^2 \langle 2\lambda^2 + 2\lambda + 1 \rangle], \quad (11.54a)$$

$$\langle |\theta_{\mu c}|^4 \rangle - \langle |\theta_{\mu c}|^2 \rangle^2 = 2J_c^2 [2 \langle (\lambda - \langle \lambda \rangle)^2 + \langle 2\lambda^2 + 2\lambda + 1 \rangle] \quad (11.54b)$$

where

$$\langle \lambda^n \rangle = \frac{\int \lambda^n [\lambda(1+\lambda)]^{\frac{1}{2}(N-3)} p_N(\lambda) \, d\lambda}{\int [\lambda(1+\lambda)]^{\frac{1}{2}(N-3)} p_N(\lambda) \, d\lambda}, \quad (11.55a)$$

$$p = \frac{\sum_\alpha J_{\alpha c} J_{\alpha c'}}{[(\sum J_{\alpha c}^2)(J_{\alpha c'}^2)]^{\frac{1}{2}}}, \quad (11.55b)$$

and the dimension N is assumed to be large.

Using eqns (11.53) and (11.54),

$$\rho^{(c,c')}_{|\theta|^2} = \frac{p^2 + 2\langle (\lambda - \langle \lambda \rangle)^2 \rangle/(1 + 2\langle \lambda(1+\lambda) \rangle)}{1 + 2\langle (\lambda - \langle \lambda \rangle)^2 \rangle/(1 + 2\langle \lambda(1+\lambda) \rangle)}. \quad (11.56)$$

Denoting the channel correlation coefficients of partial width $\Gamma_{\mu c}$ by $\rho^{(c,c')}_\Gamma$ and of the quantity $\Theta_{\mu c}$ by $\rho^{(c,c')}_\Theta$, it can be shown that

$$\rho^{(c,c')}_\Gamma = p^2, \quad (11.57)$$

$$\rho^{(c,c')}_\Theta = \langle [(1+2\lambda)^2 - \langle (1+2\lambda)^2 \rangle]^2 \rangle + \frac{2p^2 \langle 1 + 6\lambda + 14\lambda^2 + 16\lambda^3 + 8\lambda^4 \rangle}{S(\Theta_{\mu c}) \langle N_\mu^2 \rangle}. \quad (11.58)$$

We see from eqns (11.56)–(11.58) that channel–channel correlation is always positive. For the magnitude of correlation coefficients we use the weighting function (11.50) and calculate $\langle \lambda^n \rangle$ using (11.55a). Taking the value of the correlation coefficient $\rho_g^{(c,c')}$ which is the same as $\rho_{|\theta|^2}^{(c,c')}$ from the numerical calculation, we find that $p^2 = 0.087$. This gives us $\rho_\Theta^{(c,c')} = 0.32$ and $\rho_\Gamma^{(c,c')} = 0.09$ which should be compared with their numerical values of 0.20 and 0.002, respectively.

Finally we consider the resonance correlations. The resonance correlation $\rho_N^{(1)}$ is defined by

$$\rho_N^{(1)} = \frac{\langle N_\mu N_{\mu+1}\rangle - \langle N_\mu\rangle^2}{\langle N_\mu^2\rangle - \langle N_\mu\rangle^2}. \tag{11.59}$$

The resonance correlation is more difficult to calculate as it involves two different columns of the complex orthogonal matrix. The ensemble average of $\langle N_\mu N_{\mu+1}\rangle$ for large dimensions of the complex orthogonal matrix can be expressed approximately as

$$\langle N_\mu N_{\mu+1}\rangle = k^{-1} \int (2\lambda_1 + 1)(2\lambda_2 + 1)\rho_N(\lambda_1)\rho_N(\lambda_2)$$
$$\times [\lambda_1(1+\lambda_1)\lambda_2(1+\lambda_2)]^{\frac{1}{2}(N-3)}[(1+\lambda_1+\lambda_2+2\lambda_1\lambda_2)$$
$$\times (\lambda_1+\lambda_2+2\lambda_1\lambda_2)]^{\frac{1}{2}}\,d\lambda_1\,d\lambda_2 \tag{11.60}$$

where k is the same integral but without the factor $(2\lambda_1 + 1)(2\lambda_2 + 1)$. Using the form of $\rho_N(\lambda)$ given by eqn (11.50), the integral in eqn (11.60) is evaluated approximately and $\rho_N^{(1)}$ calculated using eqn (11.59). Its value turns out to be 0.2 which is somewhat bigger than its exact value.

11.4 Average value and fluctuation of various cross-sections

We shall now study the average value and the fluctuation of various cross-sections. We first define the averages of the partial cross-sections by the relations

$$\langle \sigma_c \text{ total}\rangle_{E_0} = 2\pi \lambdabar_c^2(1 - \text{Re}\langle U_{cc}^s(E, E_0)\rangle_{\text{av}}), \tag{11.61a}$$

$$\langle \sigma_{cc'}\rangle_{E_0} = 2\pi \lambdabar_c^2 \langle |\delta_{cc'} - U_{cc'}^s(E, E_0)|^2\rangle_{\text{av}} \tag{11.61b}$$

where $\lambdabar_c$ is the inverse of k_c, E_0 denotes the value of energy E at which the expectation value of the cross-section is taken, and $\langle\ \rangle_{\text{av}}$ denote the averages of the collision matrix $U_{cc'}^s$ and Re means the real part of the subsequent expression.

$$\sigma_{cc'}^{\text{direct}}(E_0) = \pi \lambdabar_c^2 |\delta_{cc'} - \langle U_{cc'}^s(E, E_0)\rangle_{\text{av}}|^2, \tag{11.62a}$$

$$\sigma_{cc}^{\text{direct}}(E_0) = \sigma_c^{\text{shape elastic}}(E_0), \tag{11.62b}$$

11.4 VALUE AND FLUCTUATION

$$\sigma_c^{\text{absorption}}(E_0) = \pi \lambda_c^2 T_c(E_0) \tag{11.63a}$$

$$= \pi \lambda_c^2 (1 - |\langle U_{cc}^s(E, E_0) \rangle_{\text{av}}|^2) \tag{11.63b}$$

$$\sigma_{cc'}^{\text{fluct}}(E_0) = \pi \lambda_c^2 (\langle |U_{cc'}^s|^2 \rangle_{\text{av}}$$

$$- |\langle U_{cc'}^s \rangle_{\text{av}}|^2) \tag{11.64}$$

where $\sigma_{cc'}^{\text{fluct}}$ is the partial fluctuation cross-section and T_c is the transmission coefficient.

We now average the collision matrix $U_{cc'}^s$ using the box weighting function. This gives

$$\langle U_{cc'}^s \rangle_{\text{av}} = U_{cc'}^0 - \frac{\pi}{D} \langle g_{\mu c} g_{\mu c'} \rangle_\mu \tag{11.65a}$$

where $\langle \rangle_\mu$ denotes ensemble average over resonance parameters. The transmission coefficient T_c is then given by

$$T_c = 1 - |U_{cc}^0|^2 + \frac{2\pi}{D} \text{Re}(U_{cc}^{0*} \langle g_{\mu c}^2 \rangle_\mu) - \frac{\pi^2}{D^2} |\langle g_{\mu c}^2 \rangle_\mu|^2. \tag{11.65b}$$

For the partial fluctuation cross-section given by (11.64) we need the ensemble average of $|U_{cc'}^s|^2$. As in Chapter 8, its evaluation will involve the contributions from the poles of $U_{cc'}^{s*}$ also. It can be shown that

$$\sigma_{cc'}^{\text{fluct}}(E_0) = \pi \lambda_c^2 \left[\frac{2\pi}{D} \left\langle \frac{|g_{\mu c}|^2 |g_{\mu c'}|^2}{\Gamma_\mu} \right\rangle_\mu - M_{cc'} \right] \tag{11.66}$$

where

$$M_{cc'} = \frac{2\pi^2}{D^2} \left\{ |\langle g_{\mu c} g_{\mu c'} \rangle_\mu|^2 \right.$$

$$\left. - \left\langle g_{\mu c} g_{\mu c'} g_{\mu c}^* g_{\mu c'}^* \cdot \Phi_0 \left(\frac{\Gamma_\mu \Gamma_\nu}{2D} \right)_{\mu \pm \nu} \right\rangle \right\}. \tag{11.67}$$

The function $\Phi_0(\Gamma/D)$ is defined by

$$\Phi_0 \left(\frac{\Gamma}{D} \right) = -\frac{iD}{\pi} \int_{-\infty}^{\infty} \frac{dE R_2(E)}{E - i\Gamma}, \tag{11.68}$$

where $R_2(E)$ is the two-level correlation function. Dyson (1962) showed that

$$DR_2(E) = 1 - [S(y)]^2 - \frac{dS(y)}{dy} \int_y^\infty S(t) \, dt \tag{11.69}$$

where $S(y) = (\sin y)/y$ and $y = \pi |E|/D$.

Expressions (11.65)–(11.67) can be simplified further by assuming no correlations between $g_{\mu c}$s

$$\langle g_{\mu c} g_{\mu c'} \rangle_\mu = \delta_{cc'} \langle g_{\mu c}^2 \rangle. \tag{11.70}$$

Noting that the function Φ_0 varies slowly, we can write

$$\langle U_{cc'}^s \rangle_{\text{av}} = U_{cc'}^0 - \tfrac{1}{2}\delta_{cc'} b_c \left\langle \frac{\Theta_{\mu c}}{N_\mu} \right\rangle_\mu \tag{11.71}$$

where

$$b_c = \frac{\langle g_{\mu c}^2 \rangle_\mu}{\langle |g_{\mu c}|^2 \rangle_\mu} \tag{11.72}$$

and

$$M_{cc'} = \delta_{cc'} \frac{2\pi^2}{D^2} |\langle g_{\mu c}^2 \rangle_\mu|^2 [1 - \Phi_0]$$

or

$$M_{cc'} = \tfrac{1}{2}\delta_{cc'} B_c \left\langle \frac{\Theta_{\mu c}}{N_\mu} \right\rangle^2 (1 - \Phi_0) \tag{11.73}$$

where

$$B_c = |b_c|^2. \tag{11.74}$$

The fluctuation cross-section

$$\sigma_{cc'}^{\text{fluct}} = \pi \lambda_c^2 \left\langle \frac{\Theta_{\mu c} \Theta_{\mu c'}}{\Theta_\mu} \right\rangle - M_{cc'}. \tag{11.75}$$

This expression simplifies further if we assume no direct reactions and small transmission coefficients. Under these conditions,

$$\sigma_{cc'}^{\text{fluct}} = \pi \lambda_c^2 \frac{T_c T_{c'}}{\sum_{c''} T_{c''}} W_{cc'} \tag{11.76}$$

where

$$W_{cc'} = \left\langle \frac{\Theta_{\mu c} \Theta_{\mu c'}}{\Theta_\mu} \right\rangle \Big/ \frac{\langle \Theta_{\mu c} \rangle \langle \Theta_{\mu c'} \rangle}{\langle \Theta_\mu \rangle}. \tag{11.77}$$

The factor $W_{cc'}$ is called the width fluctuation factor. We see that $\sigma_{cc'}^{\text{fluct}}$ without $W_{cc'}$ is just the well known Hauser–Feshbach formula (Hauser and Feshbach 1952). We now consider the mean-square fluctuation of cross-sections and for this purpose consider the total cross-section. The observable total cross-section $\sigma_\alpha^{\text{total}}$ is a sum over partial total

11.4 VALUE AND FLUCTUATION

cross-sections σ_c^{total}

$$\sigma_\alpha^{\text{total}} = 2\pi\lambda_\alpha^2 \sum_c g_c \, \text{Re}(\tau_{cc}^0 + \tau_{cc}^1) \tag{11.78}$$

where

$$g_c = \frac{(2J_c+1)}{(2I_c+1)(2g_c+1)}$$

is the usual spin statistical factor, and

$$\tau_{cc}^0 = 1 - U_{cc}^0, \quad \tau_{cc}' = -i\sum_\mu \frac{g_{\mu c}^2}{E - \varepsilon_\mu + \frac{i}{2}\Gamma_\mu}.$$

Following Ericson (1963) the mean-square fluctuation

$$F_\alpha = \langle(\sigma_\alpha^{\text{total}})^2\rangle - \langle\sigma_\alpha^{\text{total}}\rangle^2. \tag{11.79}$$

Putting in the averages of the product of two collision matrices,

$$F_\alpha = \pi^2 \lambda_\alpha^4 \sum_{c_1 c_2} \delta_{J_1 J_2} \delta_{\pi_1 \pi_2} g_{c_1}^2 2 \, \text{Re}\left\{\frac{2\pi}{D}\left\langle\frac{g_{\mu c_1}^2 g_{\mu c_2}^{*2}}{\Gamma_\mu}\right\rangle_\mu\right\}$$

$$- \frac{2\pi^2}{D^2}\left\langle g_{\mu c_1}^2 g_{\mu c_2}^{*2}\left[1 - \Phi_0\left(\frac{\Gamma_\mu + \Gamma_\nu}{2D}\right)\right]\right\rangle_{\mu \neq \nu} \tag{11.80}$$

where π_1, π_2 denote parity. Assuming a large number of channels and uncorrelated widths simplifies (11.80) and gives

$$F_\alpha = 2\pi\lambda_\alpha^2\left\{\sum_{c_1} g_{c_1}^2\left[\sigma_{c_1 c_1}^{\text{fluct}} + \sum_{c_2}(1-\delta_{c_1 c_2})\text{Re}\, b_{c_1} b_{c_2}^* \sigma_{c_1 c_2}^{\text{fluct}}\right]\right\}. \tag{11.81}$$

If we further assume that the sign of Re $b_{c_1} b_{c_2}^*$ fluctuates with c_2 then the main contribution will arise from the first term, which, in the case of nucleon scattering, can be estimated to yield

$$F_\alpha = \frac{1}{n}\langle\sigma_\alpha^{\text{total}}\rangle \sigma_{\alpha\alpha}^{\text{fluct}} \tag{11.82}$$

where n is a number of the order of the number of strongly competing channels c.

Similarly, after a fairly long calculation, we find that the mean-square fluctuation of the reaction cross-section defined by

$$F_{\alpha\alpha'} = \langle(\sigma_{\alpha\alpha'})^2\rangle - \langle\sigma_{\alpha\alpha'}\rangle^2, \quad \alpha \neq \alpha' \tag{11.83}$$

is given by

$$F_{\alpha\alpha'} = \frac{1}{nn'}\sigma_{\alpha\alpha'}^{\text{fluct}}(\langle\sigma_{\alpha\alpha'}\rangle + \sigma_{\alpha\alpha'}^{\text{direct}}) \tag{11.84}$$

where n and n' are the number of channels c competing strongly in the decay into alternatives α and α' respectively.

In Chapters 9 and 11 we discussed only the average value and fluctuation of the integrated cross-section. We now describe the average value and fluctuation of the differential cross-section.

The average differential cross-section can be written (Moldauer 1964a; Lane and Thomas 1958)

$$\langle d\sigma_{\alpha\alpha'}\rangle = \frac{\lambda_\alpha^2}{(2I+1)(2\mathcal{I}+1)} \sum_L \langle B_L(\alpha,\alpha')\rangle_{\mathrm{av}} \times P_L(\cos\theta_{\alpha'})d\Omega_\alpha \qquad (11.85)$$

where I and $\mathcal{I}$ are the spins of the two fragments belonging to α, α', $P_L(\cos\theta_{\alpha'})$ is the Legendre polynomial, and α' denotes the scattered beam. The quantities $B_L(\alpha,\alpha')$ are given by

$$B_L(\alpha,\alpha') = \sum_{c_1 c_1' c_2 c_2'} \delta_{S_1 S_2} \delta_{S_1' S_2'} \tfrac{1}{4}(-1)^{S_1 - S_1'}$$
$$\times \bar{Z}(l_1 J_2 l_2 J_2; S_1 L)\bar{Z}(l_1' J_1 l_2' J_2, S_1' L)\mathrm{Re}(\tau_{c_1 c_1'}\tau_{c_2 c_2'}^*) \qquad (11.86)$$

where the coefficients $\bar{Z}$ (Lane and Thomas 1958) are the products of Clebsch–Gordon and Racah coefficients. The summation over c denotes the summation over the corresponding values of S, l, and J. The diagonal form of the transition amplitude $\tau_{cc'}$ was introduced earlier; in general they are defined as

$$\tau_{cc'} = \tau_{cc'}^0 + \tau_{cc'}^1, \qquad (11.87)$$
$$\tau_{cc'}^0 = \delta_{cc'} + U_{cc'}^0, \qquad (11.88a)$$

and

$$\tau_{cc'}^1 = -i\sum_\mu \frac{g_{\mu c}g_{\mu c'}}{E - \varepsilon_\mu + \tfrac{i}{2}\Gamma_\mu}. \qquad (11.88b)$$

Substituting eqns (11.87) and (11.88) into (11.86) we can write

$$\langle B_L\rangle_{\mathrm{av}} = B_L^{\mathrm{direct}} + B_L^{\mathrm{fluct}} \qquad (11.89)$$

where

$$B_L^{\mathrm{direct}} = \sum_{12'} \bar{Z}_{12L}\bar{Z}_{1'2'L}\,\mathrm{Re}\{\tau_{c_1 c_1'}^0 \tau_{c_2 c_2'}^{0*}$$
$$+ \tau_{c_1 c_1'}^0 \langle \tau_{c_2 c_2'}^{1*}\rangle_{\mathrm{av}} + \langle \tau_{c_1 c_1'}^1\rangle_{\mathrm{av}}\tau_{c_2 c_2'}^0$$
$$+ \langle \tau_{c_1 c_1'}^1\rangle_{\mathrm{av}}\langle \tau_{c_2 c_2'}^{1*}\rangle_{\mathrm{av}}\} \qquad (11.90)$$

11.4 VALUE AND FLUCTUATION

and

$$B_L^{\text{fluct}} = \sum_{12'} \bar{Z}_{12L}\bar{Z}_{1'2'L} \,\text{Re}\{\langle \tau^1_{c_1c'_1}\tau^{1*}_{c_2c'_2}\rangle_{\text{av}}$$
$$- \langle \tau^1_{c_1c'_1}\rangle_{\text{av}}\langle \tau^{1*}_{c_2-c'_2}\rangle_{\text{av}}\}. \quad (11.91)$$

The notation in eqns (11.90) and (11.91) is

$$\bar{Z}_{12L} = \bar{Z}(l_1J_1l_2J_2; S_1L), \quad (11.92\text{a})$$

$$\sum_{12'} = \sum_{c_1c'_1c_2c'_2} -\delta_{S_1S_2}\delta_{S'_1S'_2}\tfrac{1}{4}(-1)^{S_1-S'_1}. \quad (11.92\text{b})$$

It is easy to see that

$$\langle \tau^1_{cc'}\rangle_{\text{av}} = \frac{\pi}{D}\langle g_{\mu c}g_{\mu c'}\rangle_\mu. \quad (11.93)$$

This can also be written as

$$\langle \tau^1_{cc'}\rangle_{\text{av}} = \tfrac{1}{2}\delta_{cc'}b_c\left\langle\frac{\Theta_{\mu c}}{N_\mu}\right\rangle, \quad (11.94)$$

assuming no correlations between $g_{\mu c}$ and $g_{\mu c'}$.

The average of the product of two τs can also be worked out as

$$\langle \tau^1_{c_1c'_1}\tau^1_{c_2c'_2}\rangle_{\text{av}} = \frac{\langle g_{\mu c_1}g_{\mu c'_1}\rangle_\mu \langle g^*_{\mu c_2}g_{\mu c'_2}\rangle_\mu}{D_1D_2}$$
$$+ \delta_{J_1J_2}\delta_{\pi_1\pi_2}\frac{2\pi}{D}\left\langle\frac{g_{\mu c_1}g_{\mu c'_1}g^*_{\mu c_2}g^*_{\mu c'_2}}{\Gamma_\mu}\right\rangle_\mu$$
$$- \frac{2\pi^2}{D^2}\langle g_{\mu c_1}g_{\mu c'_1}g^*_{\nu c_2}g^*_{\nu c'_2}\rangle\left(1 - \Phi_0\left(\frac{\Gamma_\mu+\Gamma_\nu}{2D}\right)\right)_{\mu\neq\nu}. \quad (11.95)$$

For the case of no correlations between $g_{\mu c}$ and $g_{\mu c'}$ (11.95) can be written as

$$\langle \tau'_{c_1c'_1}\tau'^{*}_{c_2c'_2}\rangle = [\delta_{c_1c_2}\delta_{c'_1c'_2}$$
$$+ (1-\delta_{c_1c'_1})\delta_{c_1c'_2}\delta_{c'_1c_2}]\left\langle\frac{\Theta_{\mu c_1}\Theta_{\mu c'_1}}{\Theta_\mu}\right\rangle_\mu$$
$$+ \delta_{c_1c'_1}\delta_{c_2c'_2}\left\{\tfrac{1}{4}b_{c_1}b^*_{c_2}\left\langle\frac{\Theta_{\mu c_1}}{N_\mu}\right\rangle_\mu\left\langle\frac{\Theta_{\nu c_2}}{N_\nu}\right\rangle_\nu\right.$$
$$\times [1 - 2\delta_{J_1J_2}\delta^{(1-\Phi_0)}_{\pi_1\pi_2}]$$
$$\left.+ \delta_{J_1J_2}\delta_{\pi_1\pi_2}\frac{2\pi}{D}\left\langle\frac{g^2_{\mu c_1}g^{*2}_{\mu c_2}}{\Gamma_\mu}\right\rangle_\mu, c_1 \neq c_2\right\}. \quad (11.96)$$

In the many-channel case of constant Γ_μ and uncorrelated $g_{\mu c}$ we may write

$$\frac{2\pi}{D}\left\langle\frac{g^2_{\mu c_1}g^{*2}_{\mu c_2}}{\Gamma_\mu}\right\rangle_\mu = b_{c_1}b^*_{c_2}\left\langle\frac{\Theta_{\mu c_1}\Theta_{\mu c_2}}{\Theta_\mu}\right\rangle. \quad (11.97)$$

Expressions (11.91), (11.94), (11.96), and (11.97) imply that B^{fluct}_L at low energies for inelastic processes is of the Hauser–Feshbach form except that T_c is replaced by $\langle\Theta_{\mu c}\rangle_\mu$ and there is an additional width fluctuation factor.

We can also define the mean-square deviation by the relation

$$\left\langle\left(\frac{d\sigma_{\alpha\alpha'}}{d\Omega_{\alpha'}}\right)^2\right\rangle - \left\langle\frac{d\sigma_{\alpha\alpha'}}{d\Omega_{\alpha\alpha'}}\right\rangle^2$$

$$= \lambda^4_\alpha[(2I+1)^2(2\mathscr{I}+1)^2]^{-1}\sum_{LK}F_{LK}(\alpha,\alpha')P_L(\cos\theta')P_K(\cos\theta') \quad (11.98)$$

where

$$F_{LK}(\alpha,\alpha') = \langle B_L(\alpha,\alpha')B_K(\alpha,\alpha')\rangle - \langle B_L(\alpha,\alpha')\rangle\langle B_K(\alpha,\alpha')\rangle. \quad (11.99)$$

After a fairly long calculation it can be shown that for non-elastic processes, $\alpha \neq \alpha'$, and in the limit of large Γ/D we can write

$$\frac{\left\langle\left(\frac{d\sigma_{\alpha\alpha'}}{d\Omega_{\alpha'}}\right)^2\right\rangle - \left\langle\frac{d\sigma_{\alpha\alpha'}}{d\Omega_{\alpha'}}\right\rangle^2}{\left\langle\frac{d\sigma_{\alpha\alpha'}}{d\Omega_{\alpha'}}\right\rangle^2} = \frac{\sum_{LK}\sum_{12'}\bar{Z}_{12L}\bar{Z}_{1'2'L}\bar{Z}_{21K}\bar{Z}_{2'1'K}\sigma^{\text{fluct}}_{c_1c'_1}\sigma^{\text{fluct}}_{c_2c'_2}P_LP_K}{\sum_{LK}\sum_{12'}\bar{Z}_{11L}\bar{Z}_{1'1'L}\bar{Z}_{22K}\bar{Z}_{2'2'K}\sigma^{\text{fluct}}_{c_1c'_1}\sigma^{\text{fluct}}_{c_2c'_2}P_LP_K}. \quad (11.100)$$

We have been discussing the average value and the fluctuation of cross-sections but it is also possible to study the probability density function of the real and imaginary parts of the scattering matrix which can be used to study, for example, the probability density function of the partial cross-section. We use the method of moments to study the distribution of the real and imaginary parts. We add here that such distributions were arrived at by Ericson (1963) using the central limit theorem and in certain other studies (Brink and Stephens 1963; Brink et al. 1964) of cross-section distribution these distributions were used as one of the basic assumptions of the theory.

We write the fluctuating part of the partial scattering amplitude as

$$U^{s\,\text{fluct}}_{cc'} = -i\sum_\mu\frac{g_{\mu c}g_{\mu c'}}{E-\varepsilon_\mu+\frac{i}{2}\Gamma_\mu}, \quad c \neq c'. \quad (11.101)$$

We first consider the moments of Re $U^{s\,\text{fluct}}_{cc'}$. The first moment of $U^{s\,\text{fluct}}_{cc'}$

11.4 VALUE AND FLUCTUATION

was calculated earlier, giving

$$\langle \operatorname{Re} U_{cc'}^{\text{s fluct}}\rangle_{\text{av}} = -\frac{\pi}{D}\langle g_{\mu c}^* g_{\mu c'}^*\rangle_\mu. \quad (11.102)$$

This is zero under the assumption that the quantities $g_{\mu c}$ are uncorrelated.

The second moment of $U_{cc'}^{\text{s fluct}}$ was also calculated while deriving expressions for the mean-square deviation, giving

$$\langle (\operatorname{Re} U_{cc'}^{\text{s fluct}})^2\rangle_{\text{av}} = \frac{\pi}{D}\left\langle \frac{|g_{\mu c}|^2 |g_{\mu c'}|^2}{\Gamma_\mu}\right\rangle$$

$$+ \frac{\pi^2}{D^2}\langle g_{\mu c}^* g_{\mu c'}^* g_{\nu c} g_{\nu c'}\rangle \Phi_0\left(\frac{\Gamma_\mu + \Gamma_\nu}{2D}\right), \quad \mu \neq \nu. \quad (11.103)$$

For the case of large Γ/D, eqn (11.103) becomes

$$\langle (\operatorname{Re} U_{cc'}^{\text{s fluct}})^2\rangle_{\text{av}} = \frac{\pi}{D}\frac{\langle |g_{\mu c}|^2 |g_{\mu c'}|^2\rangle_\mu}{\Gamma} \quad (11.104)$$

where Γ is the average width.

The third- and fourth-order moments of $\operatorname{Re} U_{cc'}^{\text{s fluct}}$ are complicated. After a fairly long calculation we find, using the assumptions used to derive eqns (11.102) and (11.104) that

$$\langle (\operatorname{Re} U_{cc'}^{\text{s fluct}})^3\rangle = -\frac{3\pi}{2D}\frac{\langle |g_{\mu c}|^2 |g_{\mu c'}|^2 \operatorname{Re}(g_{\mu c}g_{\mu c'})\rangle_\mu}{\Gamma^2} \quad (11.105)$$

$$\langle (\operatorname{Re} U_{cc'}^{\text{s fluct}})^4\rangle = \frac{3\pi^2}{D^2}\left[\frac{\langle |g_{\mu c}|^2 |g_{\mu c'}|^2\rangle_\mu}{\Gamma}\right]^2. \quad (11.106)$$

To show that the third-order moment of $\operatorname{Re} U_{cc'}^{\text{s fluct}}$ vanishes, we write the probability density function of the complex amplitude $\theta_{\mu c}$ which is proportional to $g_{\mu c}$. This probability density $P(\theta_{1c}^{\text{Re}}, \theta_{1c}^{\text{Im}}, \theta_{1c'}^{\text{Re}}, \theta_{1c'}^{\text{Im}})$ is given by

$$P(\theta_{1c}^{\text{Re}}, \theta_{1c}^{\text{Im}}, \theta_{1c'}^{\text{Re}}, \theta_{1c'}^{\text{Im}}) = \frac{(N-2)(N-3)}{4\pi^2 N^2 |\Sigma|}$$

$$\times \left[\int \rho_N(\lambda)[\lambda(1+\lambda)]^{\frac{1}{2}(N-3)}\,d\lambda\right]^{-1}$$

$$\times \int d\lambda \rho_N(\lambda)[\lambda(1+\lambda)]^{\frac{1}{2}(N-5)}$$

$$\times \left[1 - \frac{1}{N}\left[\frac{1}{1+\lambda}(\Sigma^{-1})_{11}(\theta_{1c}^{\text{Re}})^2 + 2\{(\Sigma^{-1})_{12}\theta_{1c}^{\text{Re}}\theta_{1c}^{\text{Im}}\right.\right.$$

$$+ (\Sigma^{-1})_{22}(\theta_{1c'}^{\text{Re}})^2\} + \frac{1}{\lambda}\{(\Sigma^{-1})_{11}(\theta_{1c}^{\text{Im}})^2$$

$$+ 2(\Sigma^{-1})_{12}\theta_{1c}^{\text{Im}}\theta_{1c'}^{\text{Im}} + (\Sigma^{-1})_{22}(\theta_{1c'}^{\text{Im}})^2\}\Bigg]$$

$$+ \frac{1}{(1+\lambda)N^2 |\Sigma\|} (\theta_{1c}^{\text{Re}}\theta_{1c'}^{\text{Im}} - \theta_{1c'}^{\text{Re}}\theta_{1c}^{\text{Im}})^2 \Bigg\rangle^{\frac{1}{2}(N-5)} \quad (11.107)$$

where N is the dimension of the complex orthogonal matrix and Σ^{-1} is the matrix

$$\Sigma^{-1} = \begin{pmatrix} [(1-\gamma^2)J_c]^{-1} & -\gamma(1-\gamma^2)^{-1}(J_cJ_{c'})^{\frac{1}{2}} \\ -\gamma(1-\gamma^2)^{-1}(J_cJ_{c'})^{-\frac{1}{2}}[(1-\gamma^2)J_{c'}]^{-1} \end{pmatrix} \quad (11.108)$$

where

$$J_c = \frac{1}{N}\sum_{\alpha=1}^{N} J_{\alpha c'}^2 \quad (11.109a)$$

and

$$\gamma = \frac{\sum J_{\alpha c}J_{\alpha c'}}{[(\sum J_{\alpha c}^2)(\sum J_{\alpha c'}^2)]^{\frac{1}{2}}}. \quad (11.096)$$

Taking N to be very large, we find that the assumption of uncorrelated $g_{\mu c}$ implies $\gamma = 0$, which shows that the third-order moments of Re $U_{cc'}^{\text{s fluct}}$ vanish. From eqns (11.104) and (11.106),

$$\frac{\langle(\text{Re } U_{cc'}^{\text{s fluct}})^4\rangle_{\text{av}}}{\langle(\text{Re } U_{cc'}^{\text{s fluct}})^2\rangle_{\text{av}}^2} = 3, \quad (11.110)$$

which is characteristic of Gaussian distribution.

Thus the real part of $U_{cc'}^{\text{s fluct}}$ has a Gaussian distribution with mean zero and variance

$$\frac{\pi}{D} \frac{\langle |g_{\mu c}|^2 |g_{\mu c'}|^2\rangle_\mu}{\Gamma}.$$

A similar result can be proved for the imaginary part of $U_{cc'}^{\text{s fluct}}$.

Therefore both the real and imaginary parts of $U_{cc}^{\text{s fluct}}$ have a Gaussian distribution with zero mean and the same variance.

11.5 Concluding remarks

We have given various expressions for the average and mean-square deviation of various cross-sections. We have shown that in order to derive these expressions we must first study the averages of the parameters of the statistical collision matrix. Under certain conditions the average reaction cross-section becomes the same as the Hauser–Feshbach expression. In the analysis of experimental data on compound nucleus reactions we have to choose among the expressions given in Chapters 8, 10, and 11 according to the energy of the neutron beam and the target nucleus, e.g. for low-energy neutrons and for isolated resonances we use the Hauser–Feshbach expression (Chapter 8); otherwise, we use one of the more exact expressions given in Chapters 10 or 11.

11.6 References

Brink, D. M. and Stephen, R. O. (1963). *Physics Letters* **5,** 77.
Brink, D. M., Stephen, R. O., and Tanner, N. W. (1964). *Nuclear Physics* **54,** 577.
Dyson, F. J. (1962). *Journal of Mathematical Physics* **3,** 166.
Ericson, T. (1963). *Annals of Physics, New York* **23,** 390.
Hauser, W. and Feshbach, H. (1952). *Physical Review* **87,** 366.
Lane, A. M. and Thomas, R. G. (1958). *Reviews of Modern Physics* **30,** 257.
Moldauer, P. A. (1964a). *Physical Review* **135,** B642.
Moldauer, P. A. (1964b). *Physical Review* **136,** B947.
Porter, C. E. (1965). *Statistical theories of spectra. Fluctuations.* Academic Press, New York.
Ullah, N. (1967). *Physical Review* **154,** 891, 893.
Whittaker, E. T. and Watson, G. N. (1962). *A course of modern analysis.* Cambridge University Press, New York.
Wigner, E. P. (1959). *Group theory.* Academic Press, New York.

12

CONFIGURATION INTERACTION PROBLEM

12.1 Introduction

One of the oldest problems in many-body physics is the configuration interaction problem. In the early days of atomic physics, a Hartree–Fock (H–F) calculation using a single Slater determinant was the starting point in generating a many-electron wave function. The H–F calculation generates a set of single-electron wave functions which can then be used to obtain a complete set of many-body wave functions (Nesbet 1955) by operating with the particle–hole operator on the approximate single Slater determinant. Because of the H–F property, the matrix element of the many-electron Hamiltonian vanishes between the trial wave function and the one-particle–one-hole wave functions. Since the interaction is a two-body term, the matrix elements of the many-body Hamiltonian between the trial wave function and wave functions which have more than two particles–two-holes in them vanish. Thus H–F calculation provides a very good starting point for applying configuration interaction corrections to the trial wave function. Since the number of two-particle–two-hole wave functions is quite large, we do not diagonalize the Hamiltonian using all these wave functions but use only those wave functions which interact strongly with the trial ground-state wave function, the rest of the two-particle–two-hole wave functions are taken care of by second-order perturbation theory.

When H–F theory was applied to many-nucleon systems it was found to be advantageous to use what is known as the deformed H–F wave function (Kelson and Levinson 1964). This wave function does not possess a good angular momentum. Good angular momentum wave functions are obtained by using an angular momentum projection operator. Part of the configuration interaction is taken into account this way. For the rest we have again to diagonalize the Hamiltonian or use perturbation theory.

In this chapter we would like to describe statistical methods for studying the configuration interaction problem. In § 12.2 we describe the general background to using statistical methods to attack the configuration interaction problem. § 12.3 describes a combination of statistical methods and perturbation theory. In § 12.4 we use the concepts of centroids and partial widths to find the contribution of higher configurations to the ground-state energy of the system.

12.2 The configuration interaction problem using statistical methods

From the discussion in § 12.1 it is clear that the usual matrix ensembles in which all states are treated equally cannot be used to describe the configuration interaction problem. Whether a shell model picture or the variational approach is used, it is stated that there is a dominant component in the exact wave function of the many-body system. It is the other components which mostly come from higher configurations which should be treated using the ideas of matrix ensemble theory. We first describe the formulation for the ground state of the system.

Let Ψ_0 be the true ground state of the system which has a dominant component Φ_0. We now use the other configurations and construct wave functions having the same value of J^π and T (T being the isotopic spin quantum number), e.g. by using the particle–hole operator. We assume that the set Φ_μ ($\mu = 0, 1, \ldots, N$) is orthonormal. The dimension N is usually quite large. The components Φ_μ ($\mu = 1, \ldots, N$) are treated statistically alike. The wave function Ψ_0 can then be written as

$$\Psi_0 = \lambda \left[\Phi_0 + \sum_{\mu=1}^{N} a_\mu \Phi_\mu \right] \quad (12.1)$$

where the normalization constant λ is given by

$$\lambda^2 \left[1 + \sum_{\mu=1}^{N} a_\mu^2 \right] = 1. \quad (12.2)$$

The expectation value of some operator Q in the ground state Ψ_0 can then be written as

$$Q_{00} = \lambda^2 \left[q_{00} + 2 \sum_{\mu=1}^{N} a_\mu q_{\mu 0} + \sum_{\mu=1}^{N} a_\mu^2 q_{\mu\mu} + \sum_{\mu \neq \nu} a_\mu a_\nu q_{\mu\nu} \right], \quad (12.3)$$

$$Q_{00} = (\Psi_0 | Q | \Psi_0), \quad (12.4a)$$

$$q_{\mu\nu} = (\Phi_\mu | Q | \Phi_\nu). \quad (12.4b)$$

The underlying idea of matrix ensemble theory is that the joint probability density function of the coefficients a_μ should remain invariant under any linear orthogonal transformation of the set Φ_μ ($\mu = 1, \ldots, N$). If we now consider a_μ to be the components of an N-dimensional vector, this implies that joint probability density function $P(\{a_\mu\})$ of the coefficients a_μ can be written

$$P(\{a_\mu\}) \prod_{\mu=1}^{N} da_\mu = f\left(\prod_{\mu=1}^{N} a_\mu^2 \right) \prod_{\mu=1}^{N} da_\mu \quad (12.5)$$

where the form of the function f has to be determined later using some plausible physical picture.

It should be noted here that, once the function f is known, the statistical properties of the normalization constant λ are also known using eqn (12.2).

Without knowing the explicit form of f, we can write the following general expressions for the mean and mean-square deviations of Q_{00}

$$\langle Q_{00}\rangle = q_{00} + \langle 1-\lambda^2\rangle \frac{1}{N}\sum_{\mu}(q_{\mu\mu}-q_{00}), \tag{12.6}$$

$$\langle Q_{00}^2\rangle - \langle Q_{00}\rangle^2 = [\langle a_\mu^4\lambda^4\rangle - \langle a_\mu^2\lambda^2\rangle^2]$$

$$\times \sum_\mu (q_{\mu\mu}-q_{00})^2 + [\langle a_\mu^2 a_\nu^2\lambda^4\rangle_{\mu\neq\nu} - \langle A_\mu^2\lambda^2\rangle^2]$$

$$\times \sum_{\mu\neq\nu}^N (q_{\mu\mu}-q_{00})(q_{\nu\nu}-q_{00})$$

$$+ 4\langle a_\mu^2\lambda^4\rangle \sum_\mu q_{\mu 0}^2 + 2\langle a_\mu^2 a_\nu^2\lambda^4\rangle_{\mu\neq\omega}\sum_{\mu\neq\nu} q_{\mu\mu}^2 \tag{12.7}$$

where the bracket sign $\langle\ \rangle$ denotes the ensemble average.

It is easy to show that eqns (12.6) and (12.7) reproduce the fully statistical result which is used in the discussion of gyromagnetic ratios (Porter 1965) if we treat all Φ_μ ($\mu = 0, \ldots, N$) alike and assume that Φ_μ is the diagonal representation of the operator Q. For this special case eqns (12.6) and (12.7) become

$$\langle Q_{00}\rangle = \frac{1}{N+1}\left(\sum_{\mu=0}^N q_{\mu\mu}\right), \tag{12.8}$$

$$\langle Q_{00}^2\rangle - \langle Q_{00}\rangle^2 = \frac{2}{N+3}\left[\frac{1}{N+1}\sum_{\mu=0}^N q_{\mu\mu}^2 - \left(\frac{1}{N+1}\sum_{\mu=0}^N q_{\mu\mu}\right)^2\right]. \tag{12.9}$$

We next discuss the choice of the function f. An ensemble which is known as a fixed-strength ensemble was introduced by Rosenzweig (1962) to treat all interactions as equally likely. This form seems to be the most suitable for the function f. Mathematically,

$$f\left(\sum_{\mu=1}^N a_\mu^2\right) = K^{-1}\delta\sum_{\mu=1}^N [a_\mu^2 - R^2] \tag{12.10}$$

where R is a constant and K is fixed by the normalization condition,

$$\int P(\{a_\mu\})\prod_{\mu=1}^N da_\mu = 1. \tag{12.11}$$

We note here that when the dimension N is very large, the distribution given by (12.10) becomes a Gaussian distribution.

Using eqns (12.5), (12.10), and (12.11) and the evaluation of the

12.2 THE CONFIGURATION INTERACTION PROBLEM

integrals involving the δ-function as described in Chapter 9, we can show that

$$\langle a_\mu^{2m} a_\nu^{2n} \rangle = \frac{R^{2m+2n}}{\pi} \frac{\Gamma(\tfrac{1}{2}N)}{\Gamma(\tfrac{1}{2}N+m+n)} \Gamma(\tfrac{1}{2}+m)\Gamma(\tfrac{1}{2}+n) \quad (12.12)$$

where $\mu \neq \nu$ and m and n are integers. It is obvious that the odd moments of a_μ or a_ν vanish.

Using eqns (12.2), (12.6), (12.7), and (12.12),

$$\langle Q_{00} \rangle = q_{00} + \frac{R^2}{1+R^2} \frac{1}{N} \sum_\mu (q_{\mu\mu} - q_{00}), \quad (12.13)$$

$$\langle Q_{00}^2 \rangle - \langle Q_{00} \rangle^2 = \frac{2R^4}{(1+R^2)^2} \frac{N-1}{N(N+2)}$$
$$\times \left[\frac{1}{N} \sum_\mu (q_{\mu\mu} - q_{00})^2 - \frac{1}{N(N-1)} \right.$$
$$\left. \times \sum_{\mu \neq \nu} (q_{\mu\mu} - q_{00})(q_{\nu\nu} - q_{00}) \right]$$
$$+ \frac{2R^4}{(1+R^2)^2} \frac{1}{N(N+2)} \sum_{\mu \neq \nu} q_{\mu\nu}^2 + \frac{4R^2}{(1+R^2)^2} \frac{1}{N} \sum_\mu q_{\mu 0}^2. \quad (12.14)$$

We now use eqns (12.13) and (12.14) to show why the H–F approximation provides a very good starting point for the configuration interaction problem. It is obvious that, if the dispersion of some operator Q given by (12.14) is small, the value of Q_{00} will remain close to its average value $\langle Q_{00} \rangle$. As we remarked in § 12.1, if we take Φ_0 to be a trial Slater determinant, we can construct other Φ_μs using the particle–hole operation and further, if Φ_0 is obtained by the H–F approximation, Φ_μ will be 2p–2h wave functions. Now, if we have a single-particle operator such as a magnetic dipole moment operator, $q_{\mu 0}$ in (12.14) will vanish. The first term enclosed by brackets in eqn (12.14) will always be small as it goes like N^{-1} for large N. The term $\sum_{\mu \neq \nu} q_{\mu\nu}^2$ which must be examined more carefully. For the single-particle operator it can be shown (Ullah 1971) that it goes like $N^{-\tfrac{1}{2}}$ for large N. Thus the dispersion in the H–F approximation becomes quite small so that it is a good zero-order approximation for calculating various properties of a many-body system.

We next consider an example which can be used to check some of the predictions of the statistical model. We recall that eqn (12.6) was derived by assuming an invariant distribution for the coefficients a_μ. How far this distribution is justified can be checked by calculating $\langle Q_{00} \rangle$ and q_{00} for some known nucleus and the operator Q. For this purpose we consider the ground state of the ^{16}O nucleus. The dominant wave

Table 12.1 Single-particle energies for the correlated and uncorrelated wave functions for the ^{16}O nucleus. The overlap $|\langle \Phi_0 | \Psi_0 \rangle|^2 = 0.50$

	Uncorrelated Φ_0		Correlated Ψ_0	
Single-particle state	Occupancy of single-particle state	Single-particle energy (MeV)	Occupancy of single-particle state	Single-particle energy (MeV)
$1s_{\frac{1}{2}}$	1	−45.78	1.00	−44.68
$1p_{\frac{3}{2}}$	1	−21.63	0.940	−20.70
$1p_{\frac{1}{2}}$	1	−17.92	0.880	−17.36
$1d_{\frac{5}{2}}$	0	−0.01	0.035	−0.64
$2s_{\frac{1}{2}}$	0	−0.71	0.035	−0.64
$1d_{\frac{3}{2}}$	0	4.03	0.045	4.15
$2p_{\frac{3}{2}}$	0	11.54	0	11.82
$2p_{\frac{1}{2}}$	0	12.87	0	13.15
$1f_{\frac{7}{2}}$	0	15.00	0	15.15
$1f_{\frac{5}{2}}$	0	19.85	0	19.79

function Φ_0 is approximated by the closed shell $(1s_{\frac{1}{2}})^4 (1p_{\frac{3}{2}})^8 (1p_{\frac{1}{2}})^4$. The improved wave function Ψ_0 is taken to be the one which includes all $2\hbar\omega$ excitations. The single-particle operator Q is chosen to be $\{a_\nu, [H, a_\nu^\dagger]\}$, where $a_\nu^\dagger$ creates a single nucleon in orbit ν. The $\{\ \}$ denotes one-half of the sum of the double commutator $[a_\nu, [H, a_\nu^\dagger]]$ and its Hermitian conjugate. The vacuum expectation value of Q gives the single-particle energies ε_ν. In Table 12.1 we show the values of ε_ν for Φ_0 as well as Ψ_0. In the same table we also show the occupancies of the single-particle states calculated by Wong (1968). We find from Table 12.1 that, even though the wave function Ψ_0 has changed quite a bit, the single-particle energies have not changed by more than a few per cent, thus justifying the vanishing of the average value of the coefficient a_μ.

The same type of behaviour is seen when the root-mean-square radius $\langle r^2 \rangle$ is calculated. Table 12.2 shows the values of $\langle r^2 \rangle$ for the nuclei ^{16}O, ^{40}Ca, and ^{208}Pb obtained by Agassi, Gillet, and Lumbroso (1969) using uncorrelated and correlated ground states for these nuclei.

We also find the same behaviour if we consider the density distribution in the nucleus ^{40}Ca. This is shown in Fig. 12.1 for the density distribution obtained by Brown and Jacob (1963) for the correlated and uncorrelated wave functions for the nucleus ^{40}Ca; Fig. 12.2 shows the density distribution obtained by Agassi et al. (1969).

Before we leave this section we remark that we could also study the

12.2 THE CONFIGURATION INTERACTION PROBLEM

Table 12.2 Root-mean-square radii $\langle r^2 \rangle$ for the uncorrelated and correlated wave functions

Nucleus	Root-mean-square radius $\langle r^2 \rangle$ (fm)	
	Uncorrelated wave function	Correlated wave function
^{16}O	6.92	7.10
^{40}Ca	10.34	10.84
^{208}Pb	35.68	35.93

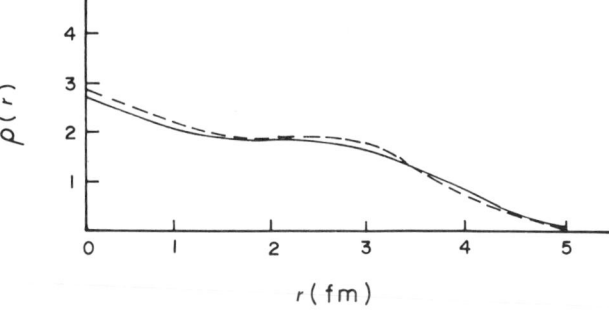

Fig. 12.1 Density distribution $\rho(r)$ for the ground state of ^{40}Ca (Brown and Jacob 1963) using correlated (——) and uncorrelated (– – –) wave functions. $\int_0^\infty \rho(r)r^2\,dr = 40$.

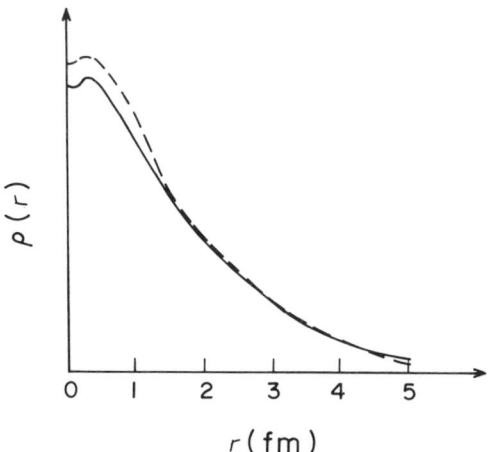

Fig. 12.2 $\rho(r)$ for the ground state of ^{40}Ca (Agassi *et al.* 1969). $\rho(r)$ in arbitrary units.

probability density function of Q_{00} given by (12.3). In general, it will be quite a difficult task to obtain an expression. Here, we shall derive the moment generating function (Kendall 1959) for Q_{00} when $N=2$, using the distribution given by eqn (12.10).

We rewrite eqn (12.3) using eqn (12.10)

$$(1+R^2)^{-1}\left(Q_{00} - \frac{q_{00}}{1+R^2}\right) = 2\sum_{\mu=1}^{2} a_\mu q_{\mu 0} + \sum_{\mu=1}^{2} a_\mu^2 q_{\mu\mu} + a_1 a_2 (q_{12} + q_{21}). \quad (12.15)$$

Writing X for $(1+R^2)^{-1}\{Q_{00} - (q_{00}/1+R^2)\}$,

$$X = 2\sum_{\mu=1}^{2} a_\mu q_{\mu 0} + \sum_{\mu=1}^{2} a_\mu^2 q_{\mu\mu} + 2q a_1 a_2 \quad (12.16)$$

where $2q = q_{12} + q_{21}$.

The moment generating function is obtained by multiplying both sides by λ, raising to power n, dividing by $n!$, and summing over n; this gives

$$\exp(\lambda X) = \exp \lambda (2\sum a_\mu q_{\mu 0} + \sum a_\mu^2 q_{\mu\mu} + 2q a_1 a_2). \quad (12.17)$$

Taking the ensemble average using eqn (12.10),

$$\langle \exp \lambda X \rangle = K^{-1} \int \exp\{\lambda[2\sum a_\mu q_{\mu 0} + \sum a_\mu^2 q_{\mu\mu} + 2q a_1 a_2]\}$$
$$\times \delta(\sum a_\mu^2 - R^2) \prod da_\mu. \quad (12.18)$$

By replacing a by Ra_μ we can rewrite (12.18) as

$$\langle (\exp \lambda X) \rangle = K^{-1} \int \exp\{\lambda[2R(\sum a_\mu q_{\mu 0})$$
$$+ R^2(\sum a_\mu^2 q_{\mu\mu} + 2q a_1 a_2)]\} \delta(\sum a_\mu^2 - 1) \prod da_\mu. \quad (12.19)$$

We now apply an orthogonal transformation to diagonalize the matrix

$$\begin{pmatrix} q_{11} & q \\ q & q_{22} \end{pmatrix}.$$

12.3 PERTURBATIVE STATISTICAL METHOD

Let d_1, d_2 be the eigenvalues; then eqn (12.19) can be written

$$\langle \exp \lambda X \rangle = K^{-1} \int \exp\{\lambda[2R(p_1 b_1 + p_2 b_2)$$
$$+ R^2(d_1 b_1^2 + d_2 b_2^2)]\} \delta(\sum b_\mu^2 - 1) \prod db_\mu \quad (12.20)$$

where $b = Oa$, O being the orthogonal matrix and

$$p_1 = (O_{11} q_{10} + O_{12} q_{20}), \quad (12.21a)$$
$$p_2 = (O_{21} q_{10} + O_{22} q_{20}). \quad (12.21b)$$

Introducing polar coordinates we get an integral representation for the moment generating function

$$\langle \exp \lambda X \rangle = (2K)^{-1} \int_{\theta=0}^{2\pi} \exp\{\lambda[2R(p_1 \cos\theta + p_2 \sin\theta)$$
$$+ R^2((d_1 - d_2)\cos^2\theta + d_2)]\} \, d\theta. \quad (12.22)$$

It can easily be checked that this expression reproduces the first two moments of X given by eqns (12.13) and (12.14) when $N = 2$.

12.3 Perturbative statistical method

In the last section we wrote the ground-state wave function as having a dominant component and the rest of the configuration interaction wave functions were treated as equally likely. To make the statistical model of the configuration interaction problem more realistic we shall now put in the additional requirement that configurations far away in energy from the dominant component make smaller contributions than the ones which are nearer to it as one finds from first-order perturbation theory. This will be done by rewriting the coefficients with proper energy denominators. We shall now also consider the excited states as well as the ground state.

Let Φ_μ be a set of N sets having a definite J^π and T obtained from unperturbed shell model configurations. Let Ψ_α and Ψ_β be the exact wave functions corresponding to states α and β and let Φ_α and Φ_β be the corresponding dominant components in Ψ_α and Ψ_β. We can then write

$$\Psi_\alpha = \lambda_\alpha \left[\Phi_\alpha + \sum_{\mu \neq \alpha}^{N} a_{\alpha\mu} \Phi_\mu \right], \quad (12.23)$$

$$\Psi_\beta = \lambda_\beta \left[\Phi_\beta + \sum_{\mu \neq \beta}^{N} a_{\beta\mu} \Phi_\mu \right] \quad (12.24)$$

where λ_α and λ_β are normalization constants such that

$$\lambda_\alpha^2 \left[1 + \sum_{\mu \neq \alpha}^N a_{\alpha\mu}^2\right] = 1; \qquad (12.25)$$

a similar expression can be written for λ_β.

The matrix element of an operator Q between Ψ_α and Ψ_β can be written

$$Q_{\alpha\beta} = q_{\alpha\beta} + \lambda_\alpha \lambda_\beta \left[\sum_{\mu \neq \beta} a_{\beta\mu} q_{\alpha\mu} + \sum_{\mu \neq \alpha} a_{\alpha\mu} q_{\mu\beta} + \sum_{\substack{\mu \neq \alpha \\ \nu \neq \beta}} a_{\alpha\mu} a_{\beta\nu} q_{\beta\nu} a_{\mu\beta} \right.$$
$$\left. - \frac{(1 - \lambda_\alpha \lambda_\beta)}{\lambda_\alpha \lambda_\beta} q_{\alpha\beta} \right], \quad (12.26)$$

where $q_{\alpha\beta}$ is the matrix element of Q between the dominant components Φ_α and Φ_β. In a purely statistical treatment the coefficients $a_{\alpha\mu}$ are treated as statistically alike. For an improved approximation we write

$$a_{\alpha\mu} = -\frac{A_{\alpha\mu}}{\varepsilon_\mu - \varepsilon_\alpha} \qquad (12.27)$$

where ε_μ and ε_α are the unperturbed energies of the wave functions Φ_μ and Φ_α respectively. It is the coefficients $A_{\alpha\mu}$ which will now be treated using the statistical approximation. Further we do not assume that the statistical average of the coefficients $A_{\alpha\mu}$ vanishes but denote it by A_α

$$A_\alpha = \langle A_{\alpha\mu} \rangle_{\text{stat}}. \qquad (12.28)$$

This of course will introduce more unknown parameters into the theory which will have to be determined by comparing the calculated quantities with their experimental values.

Taking the average of eqn (12.26) we can write for $\alpha \neq \beta$,

$$\langle Q_{\alpha\beta} \rangle_{\text{stat}} = q_{\alpha\beta} + \lambda_\alpha \lambda_\beta \left[-A_\beta \sum_{\mu \neq \beta} \frac{q_{\alpha\mu}}{\varepsilon_\mu - \varepsilon_\beta} - A_\alpha \sum_{\mu \neq \alpha} \frac{q_{\mu\nu}}{\varepsilon_\mu - \varepsilon_\alpha} \right.$$
$$\left. + A_\alpha A_\beta \sum_{\substack{\mu \neq \alpha \\ \nu \neq \beta}} \frac{q_{\mu\nu}}{(\varepsilon_\mu - \varepsilon_\alpha)(\varepsilon_\nu - \varepsilon_\beta)} - \frac{(1 - \lambda_\alpha \lambda_\beta)}{\lambda_\alpha \lambda_\beta} q_{\alpha\beta} \right]. \quad (12.29)$$

In writing eqn (12.29) we have neglected the weak dependence of the normalization constant on the coefficients $a_{\alpha\beta}$ and, since we have assumed $\alpha \neq \beta$, we have taken

$$\langle A_{\alpha\mu} A_{\beta\mu} \rangle_{\text{stat}} = A_\alpha A_\beta,$$

i.e. no correlation between the α and β states.

12.3 PERTURBATIVE STATISTICAL METHOD

Averaging of the normalization relation (12.25) gives

$$\lambda_\alpha^2 \left[1 + \langle A_{\alpha\mu}^2 \rangle_{\text{stat}} \sum_{\mu \neq \alpha}^N \frac{1}{(\varepsilon_\mu - \varepsilon_\alpha)^2} \right] = 1. \tag{12.30}$$

We now consider the average value of the diagonal element $Q_{\alpha\alpha}$. For this case, it is convenient to separate the third term in the square brackets in eqn (12.26) into two terms

$$\sum_{\mu \neq \alpha} a_{\alpha\mu}^2 q_{\mu\mu} + \sum_{\substack{\mu \neq \alpha \\ \nu \neq \alpha \\ \mu \neq \nu}} a_{\alpha\mu} a_{\alpha\nu} q_{\mu\nu}.$$

Since $\mu \neq \nu$, in the second of these terms, we replace $\langle a_{\alpha\mu} a_{\alpha\nu} \rangle_{\text{stat}}$ by $\langle a_{\alpha\mu} \rangle^2$. Using eqns (12.26), (12.27) and (12.30),

$$\langle Q_{\alpha\alpha} \rangle_{\text{stat}} = q_{\alpha\alpha} + \lambda_\alpha^2 \left[-2 A_\alpha \sum_{\mu \neq \alpha} \frac{q_{\alpha\mu}}{\varepsilon_\mu - \varepsilon_\alpha} + A_\alpha^2 \sum_{\substack{\mu \neq \alpha \\ \nu \neq \alpha \\ \mu \neq \nu}} \frac{q_{\mu\nu}}{(\varepsilon_\mu - \varepsilon_\alpha)(\varepsilon_\nu - \varepsilon_\alpha)} \right.$$

$$\left. + (1 - \lambda_\alpha^2) \left(\sum_{\mu \neq \alpha} \frac{q_{\mu\mu}}{(\varepsilon_\mu - \varepsilon_\alpha)^2} \right) \left(\sum_{\mu \neq \alpha} \frac{1}{(\varepsilon_\mu - \varepsilon_\alpha)^2} \right)^{-1} - q_{\alpha\alpha} \right]. \tag{12.31}$$

We note from eqn (12.31) that, if all the energy denominators $(\varepsilon_\mu - \varepsilon_\alpha)^{-1}$ and $(\varepsilon_\nu - \varepsilon_\alpha)^{-1}$ are taken as unity and A_α is taken to be zero, we recover eqn (12.6) derived in § 12.1.

In deriving eqns (12.29)–(12.31) we have assumed that Ψ_α and Ψ_β belong to the same values of J^π and T and hence can be expanded in the same set of wave functions Φ_μ. We can also generalize the result to the case where Q connects states of different J^π and T. Assuming that Ψ_α can be expanded in terms of Φ_μ ($\mu = 1, 2, \ldots, N_\alpha$) and Ψ_β can be expanded in the set of wave functions χ_ν ($\nu = 1, \ldots, N_\beta$), we get

$$\langle Q_{\alpha\beta} \rangle_{\text{stat}} = q_{\alpha\beta} + \lambda_\alpha \lambda_\beta \left[-B_\beta \sum_{\mu \neq \beta}^{N_\beta} \frac{q_{\alpha\mu}}{e_\mu - e_\beta} \right.$$

$$\left. - A_\alpha \sum_{\mu \neq \alpha}^{N_\alpha} \frac{q_{\mu\beta}}{\varepsilon_\mu - \varepsilon_\alpha} + A_\alpha B_\beta \sum_{\substack{\mu \neq \alpha \\ \nu \neq \beta}} \frac{q_{\mu\beta}}{(\varepsilon_\mu - \varepsilon_\alpha)(e_\nu - e_\beta)} \right]$$

$$- \frac{(1 - \lambda_\alpha \lambda_\beta)}{\lambda_\alpha \lambda_\beta} q_{\alpha\beta}. \tag{12.32}$$

In (12.32) the energies e_ν refer to the unperturbed energies of the states χ_ν and B_β represents the statistical average $\langle B_{\beta\mu} \rangle_{\text{stat}}$. The expansion coefficients $b_{\beta\mu}$ of Ψ_β in terms of χ_μ are given by

$$b_{\beta\mu} = -\frac{B_{\beta\mu}}{e_\mu - e_\beta} \tag{12.33}$$

and the normalization constant λ_β is given by

$$\lambda_\beta^2 \left[1 + \sum_{u \neq \beta}^{N_\beta} b_{\beta\mu}^2 \right] = 1. \tag{12.34}$$

We shall now apply the above formulation to the configuration interaction for the ^{14}N nucleus. Usually the statistical models are applied to heavy nuclei but due to a very large number of interacting configurations they can also be applied to a light nucleus like ^{14}N.

The ground state of ^{14}N has $J^\pi = 1^+$, $T = 0$. The dominant configuration is the one in which there are two holes in the 1p shell. Our results will depend on the choice of the dominant wave function Φ_0 which is obtained from a 1p shell configuration having two holes in it. It is written as

$$\Phi_0 = x_1(p_{\frac{3}{2}})^{-2} + x_2(p_{\frac{3}{2}})^{-1}(p_{\frac{1}{2}})^{-1} + x_3(p_{\frac{1}{2}})^{-2} \tag{12.35}$$

where x_1, x_2, x_3 are suitable coefficients.

We shall discuss the following three cases:

1. The set of x_i obtained by exact diagonalization of the Rosenfeld interaction.
2. The set given by the calculations of Cohen and Kurath (1965).
3. An empirical set obtained by using the electric quadrupole and magnetic dipole moments. Instead of using quadrupole and dipole moment we shall use their reduced matrix elements defined by

$$\langle \Psi | |E_2| |\Psi \rangle = \frac{5}{2}\left(\frac{3}{2\pi}\right)^{\frac{1}{2}} \langle \Psi | Q_{20} |\Psi \rangle, \tag{12.36a}$$

$$\langle \Psi | |M_1| |\Psi \rangle = \frac{3}{(2\pi)^{\frac{1}{2}}} \langle \Psi | \mu |\Psi \rangle. \tag{12.36b}$$

The two independent sets of single-particle energies shown in Table 12.3 are used for the calculation of unperturbed energies. The set in the second column of Table 12.3 has been obtained from the spectrum of ^{13}C while that in the third column is the set of effective single-particle energies for ^{16}O used in shell model calculations (Wong 1966).

We first take up the exact shell model calculation. The statistical model was tested by carrying out an exact shell model calculation for the ground state of ^{14}N using all the configurations having either two holes in a 1p shell or four holes in a 1p shell and two particles in a 2s–1d shell. In order to prevent the configuration matrix from being too large, the four-hole space is restricted so that there cannot be more than two holes in the $1p_{\frac{3}{2}}$ shell alone. The calculations were done for an oscillator

12.3 PERTURBATIVE STATISTICAL METHOD

Table 12.3 Values of the single particle energies deduced from the ^{13}C spectrum and from shell model calculations for ^{16}O

Single-particle state	Single-particle energies (MeV)	
	^{13}C spectrum	Shell model calculations for ^{16}O
$1p_{3/2}$	−10.95	−21.78
$1p_{1/2}$	−4.95	−15.60
$1d_{5/2}$	−1.10	−4.15
$2s_{1/2}$	−1.86	−3.28
$1d_{3/2}$	3.39	0.93

parameter value of $v = 0.398$ fm^{-2} and a Rosenfeld mixture of 40 MeV; the single-particle energies were taken from the second column of Table 12.3. The size of the matrix turned out to be 183×183. The values of x_i were found to be

$$x_1 = 0.1539, \qquad x_2 = 0.1042, \qquad x_3 = 0.9826$$

which gave $\lambda^2 = 0.802$.

Using (12.31) the value of $\langle Q_{00} \rangle_{\text{stat}}$ was calculated. In Table 12.4 we give the values of the reduced matrix elements obtained using: (1) exact shell model diagonalization; (2) shell model plus ordinary perturbation theory; (3) the statistical model; and (4) dominant component Φ_0 only.

We next consider Φ_0 obtained using Cohen and Kurath coefficients which have the values

$$x_1 = 0.0353, \qquad x_2 = 0.2709, \qquad x_3 = 0.9620.$$

Table 12.4 Values of M_1 and E_2 obtained from experimental data, and using the exact shell model, the shell model and perturbation theory, the statistical model, and the dominant component Φ_0

Experiment	M_1 (nm)	E_2 (efm^2)
Shell model (exact)	0.4787	1.2265
Shell model (perturbation theory)	0.4217	0.3830
Statistical model	0.4219	0.3727
Dominant component Φ_0	0.4196	0.3964

Table 12.5 Values of $\langle r^2 \rangle$, M_1, and E_2 using the Cohen and Kurath ground-state wave function and a statistical model

Parameters	Cohen and Kurath ground state	Ψ_0	Experiment
$\langle r^2 \rangle$ (fm^2)	5.8270	5.8890	6.1504
M_1 (nm)	0.3917	0.4287	0.4787
E_2 (efm^2)	1.2521	1.1115	1.2265

In this case our goal is to treat four-hole–two-particle configurations on the statistical basis discussed earlier and predict the degree of configuration mixing. The independent statistical parameters λ_0 and A_0 are obtained by carrying out a best fit of the parameters $\langle r^2 \rangle$, M_1, and E_2 with their experimental values. The root-mean-square radius $\langle r^2 \rangle$ is taken to be 6.1504 fm^2 as given by Wilkinson and Mafethe (1966). The value of the oscillator parameter is taken to be 0.38 fm^{-2} and the single-particle energies listed in the second column of Table 12.3 are used.

Table 12.5 compares the values obtained using the statistical model with the dominant Φ_0 given by Cohen and Kurath. In this table we have given the values of $\langle r^2 \rangle$, M_1, and E_2 obtained using only the Cohen and Kurath ground state as well as experimental values. It is interesting to note that the values of $\langle r^2 \rangle$, M_1, and E_2 using Ψ_0 have improved in the right direction. The value of λ^2 turns out to be 0.88. Thus the statistical model predicts a 12 per cent admixture of 2s–1d shell configurations in the ground state of ^{14}N.

We shall now consider the third case in which the dominant component Φ_0 is obtained by a best fit to the experimental values of $\langle r^2 \rangle$, M_1, and E_2. This gives the coefficients x_i,

$$x_1 = -0.3287, \quad x_2 = 0.2917, \quad x_3 = 0.8983.$$

In this calculation the value of the harmonic oscillator constant is taken to be the same as was used in the second case. We now use eqn (12.31) and redo the calculation as in the second case with the new set of the values x_i. This gives us $\lambda_0^2 = 0.91$ and thus we predict an admixture of higher configurations of about 10 per cent. Table 12.6 gives the values of $\langle r^2 \rangle$, M_1, and E_2 for this case. Again we find that the statistical model has given slightly improved values of $\langle r^2 \rangle$ and M_1 but not E_2.

12.4 Probability density function of the lowest eigenvalue

The basic idea of the shell model is to diagonalize the Hamiltonian in some truncated space which is obtained by retaining some of the

12.4 PROBABILITY DENSITY FUNCTION

Table 12.6 Values of $\langle r^2 \rangle$, M_1, and E_2 using a best fit and a statistical model

Parameters	Best fit Φ_0	Statistical model Ψ_0
$\langle r^2 \rangle$ (fm)	5.8270	5.8734
M_1 (nm)	0.4785	0.4993
E_2 (efm^2)	1.2291	1.1237

configurations and leaving out the rest. It is of great interest to know how large will be contribution, e.g. to the lowest eigenvalue, if the truncated space is enlarged further. To investigate such problems, the total spectroscopic space, which is defined as the space obtained by distributing a number of nucleons among a given set of single-particle states, is divided into smaller spaces. These subspaces can be defined, e.g. by specifying particles and holes in various single-particle orbits having given values of orbital angular momentum l and total angular momentum j. Since the dimensions of these subspaces are small, the Hamiltonian is exactly diagonalized in them. The statistical methods can now be used to study interactions between these subspaces.

We consider the problem of two subspaces and first derive a joint distribution of the eigenvalues for such a problem.

Let us consider a Hamiltonian matrix which has fixed diagonal elements denoted by λ_μ ($\mu = 1, 2, \ldots, N$) where the only non-zero off-diagonal elements $H_{1\nu} = H_{\nu 1}$ ($\nu \neq 1$) have a Gaussian distribution with the same dispersion σ. We are interested in deriving the joint probability density function of the eigenvalues E_μ of this Hamiltonian matrix.

We start from the identity in E which connects the eigenvalues E_μ with the elements λ_μ and $H_{1\mu}$,

$$\prod_{\mu=1}^{N} (E - E_\mu) = \prod_{\mu=1}^{N} (E - \lambda_\mu) - \sum_{\nu=2}^{N} H_{1\nu}^2 \prod_{\mu \neq 1, \nu} (E - \lambda). \quad (12.37)$$

Equating various powers of E in identity (12.37), we obtain the relations

$$\sum_{\mu=1}^{N} E_\mu = \sum_{\mu=1}^{N} \lambda_\mu, \quad (12.38a)$$

$$\sum_{\mu<\nu} E_\mu E_\nu = \sum_{\mu<\nu} \lambda_\mu \lambda_\nu - \sum_{\mu=2}^{N} H_{1\mu}^2, \quad (12.38b)$$

$$\vdots \qquad \vdots$$

$$\prod_{\mu=1}^{N} E_\mu = \prod_{\mu=1}^{N} \lambda_\mu - \sum_{\nu=2}^{N} H_{1\mu}^2 \left(\prod_{\mu \neq 1, \nu} \lambda_\theta \right). \quad (12.38c)$$

CONFIGURATION INTERACTION PROBLEM

We now introduce a set of variables v_μ defined by

$$v_1 = \sum_{\mu<\nu}(\lambda_\mu\lambda_\nu - E_\mu E_\nu), \tag{12.39a}$$

$$v_2 = \sum (\lambda_\mu\lambda_\nu\lambda_\alpha - E_\mu E_\nu E_\alpha), \tag{12.39b}$$

$$v_{N-1} = \prod_{\mu=1}^{N}\lambda_\mu - \prod_{\mu=1}^{N}E_\mu, \tag{12.39c}$$

and derive the joint distribution function of these variables. In relation (12.39b) the summation indices μ, ν and α are all different.

From eqns (12.38) and (12.39) we can write the matrix relation

$$\begin{pmatrix} 1 & \cdots & 1 & \cdots & 1 \\ \sum_{\alpha\neq 1,2}\lambda_\alpha & \cdots & \sum_{\alpha\neq 1,3}\lambda_\alpha & \cdots & \sum_{\alpha\neq 1,N}\lambda_\alpha \\ \vdots & & \vdots & & \vdots \\ \prod_{\alpha\neq 1,2}\lambda_\alpha & \cdots & \prod_{\alpha\neq 1,3}\lambda_\alpha & \cdots & \prod_{\alpha\neq 1,N}\lambda_\alpha \end{pmatrix} \begin{pmatrix} H_{12}^2 \\ H_{13}^2 \\ \vdots \\ H_{1N}^2 \end{pmatrix} = \begin{pmatrix} v_1 \\ v_2 \\ \vdots \\ v_{N-1} \end{pmatrix} \tag{12.40}$$

between the variables v_μ and the squares of the off-diagonal elements $H_{1\mu}^2$.

Since each $H_{1\mu}$ has a Gaussian distribution, the joint probability of $H_{1\mu}$ is given by

$$P(\{H_{1\mu}\})\prod_{\mu\neq 1}\mathrm{d}H_{1\mu} = K\exp\left(-\frac{1}{2\sigma^2}\left(\sum_{\mu\neq 1}H_{1\mu}^2\right)\right)\prod_{\mu\neq 1}\mathrm{d}H_{1\mu} \tag{12.41}$$

where K is the normalization constant.

We now use the well known method (Anderson 1958) of finding the distribution of one set of variables which are related to another set by a transformation given by eqn (12.40). Thus the joint probability density function of v_μ is

$$P(\{v_\mu\})\prod_{\mu=1}^{N-1}\mathrm{d}v_\mu = K\left(\left|\prod_{\nu=1}^{N-1}\prod_{\alpha=1}^{N-1}M_{\nu\alpha}^{-1}v_\alpha\right|\right)^{-\frac{1}{2}}\exp\left(-\frac{v_1}{2\sigma^2}\right)\prod_{\mu=1}^{N-1}\mathrm{d}v_\mu \tag{12.42}$$

where K is the new normalization constant. M^{-1} is the inverse of the matrix in eqn (12.40) where

$$M^{-1} = \begin{pmatrix} \dfrac{\lambda_2^{N-2}}{\prod_{\mu\neq 1,2}(\lambda_2-\lambda_\mu)} & \cdots & \dfrac{(-1)^N}{\prod_{\mu\neq 1,2}(\lambda_2-\lambda_\mu)} \\ \vdots & & \vdots \\ \dfrac{\lambda_N^{N-2}}{\prod_{\mu\neq N}(\lambda_N-\lambda_\mu)} & \cdots & \dfrac{(-1)^N}{\prod_{\mu\neq 1,N}(\lambda_N-\lambda_\mu)} \end{pmatrix} \tag{12.43}$$

12.4 PROBABILITY DENSITY FUNCTION

We now transform the probability density function from the set of variables v_μ to E_μ using eqns (12.39) and (12.42). This gives

$$P(\{E_\mu\}) \prod_{\mu=1}^{N} dE_\mu = K\delta \sum_{\mu=1}^{N} (E_\mu - \lambda_\mu)$$

$$\times \exp\left(-\frac{1}{4\sigma^2} \sum_{\mu=1}^{N} E_\mu^2\right)\left(\prod_{\mu<\nu} |E_\mu - E_\nu|\right)$$

$$\times \left(\left|\prod_{\nu=2}^{N} \prod_{\mu=1}^{N} (\lambda_\nu - E_\mu)\right|\right)^{-\frac{1}{2}} \prod_{\mu=1}^{N} dE_\mu. \quad (12.44)$$

In writing eqns (12.43) and (12.44) we have assumed that λ_μ ($\mu = 2, \ldots, N$) are unequal. When some of these become equal, it implies that some of the eigenvalues are independent of $H_{1\mu}$ and are equal to those λ_μ which are equal. This introduces additional δ-functions in eqn (12.44) and those values of μ must be excluded from summations and products in eqn (12.44).

We now consider the distribution of the lowest eigenvalue. First consider the case for $N = 2$. The probability density function of the lowest eigenvalue E is obtained by integrating out E_2 in (12.44) from E_1 to ∞. It is given by

$$P(E) \, dE = (2\pi\sigma^2)^{-\frac{1}{2}}[(\lambda_1 - E)(\lambda_2 - E)]^{-\frac{1}{2}}(\lambda_1 + \lambda_2 - 2E)$$

$$\times \exp\left(-\frac{1}{2\sigma^2}(\lambda_1 - E)(\lambda_2 - E)\right) dE \quad (12.45)$$

where the range of E is determined from the condition that

$$(\lambda_1 - E)(\lambda_2 - E) \geq 0.$$

We can now work out the two most important characteristics of a distribution, namely the mean deviation, $\langle E \rangle$ and the mean-square deviation, $\langle E^2 \rangle - \langle E \rangle^2$. Assuming $\lambda_1 < \lambda_2$,

$$\langle E \rangle = \tfrac{1}{2}(\lambda_1 + \lambda_2) - (8(2\pi)^{\frac{1}{2}}\sigma)^{-1}(\lambda_1 - \lambda_2)^2$$
$$\times [\exp(16\sigma^2)^{-1}(\lambda_1 - \lambda_2)^2][K_0((16\sigma^2)^{-1}(\lambda_1 - \lambda_2)^2)$$
$$+ K_1((16\sigma^2)^{-1}(\lambda_1 - \lambda_2)^2)] \quad (12.46a)$$

and

$$\langle E^2 \rangle - \langle E \rangle^2 = \tfrac{1}{4}(\lambda_1 - \lambda_2)^2 + \sigma^2 - (128\pi\sigma^2)^{-1}$$
$$\times (\lambda_1 - \lambda_2)^4 \exp(8\sigma^2)^{-1}(\lambda_1 - \lambda_2)^2$$
$$\times K_0((16\sigma^2)^{-1}(\lambda_1 - \lambda_2)^2) + K_1((16\sigma^2)^{-1}(\lambda_1 - \lambda_2)^2) \quad (12.46b)$$

where K_0 and K_1 are modified Bessel functions (Abramowitz and Stegun 1965).

For the N-dimensional case, it is difficult to work out the probability density function of the lowest eigenvalue in general. However if we make the further approximation of replacing each λ_μ ($\lambda = 2, \ldots, N$) by its centroid (French and Ratcliff 1971) denoted by λ_2 then the probability density function of the lowest eigenvalue is

$$P(E)\,dE = C_N[\lambda_1 + \lambda_2 - 2E][(\lambda_1 - E)(\lambda_2 - E)]^{(N-3)/2}$$
$$\times [\exp - (2\sigma^2)^{-1}(\lambda_1 - E)(\lambda_2 - E)]\,dE. \quad (12.47)$$

The range of E is again given by the condition $(\lambda_1 - E)(\lambda_2 - E) \geq 0$. The normalization constant C_N is given by

$$C_N = (2\sigma^2)^{-(N-1)/2}[\Gamma(\tfrac{1}{2}(N-1))]^{-1}. \quad (12.48)$$

It becomes more difficult to work out the mean and mean-square deviation; they can be obtained from the relations

$$\langle (E - \tfrac{1}{2}(\lambda_1 + \lambda_2)) \rangle = -(8(2)^{\frac{1}{2}}\sigma)^{-1}(\lambda_1 + \lambda_2)^2$$
$$\times [\Gamma(\tfrac{1}{2}(2m+1))]^{-1}\left[\left(-\frac{\partial}{\partial \alpha}\right)^m (\exp(\tfrac{1}{2}\xi\alpha)\right.$$
$$\left.\times K_0(\tfrac{1}{2}\xi\alpha) + K_1(\tfrac{1}{2}\xi\alpha))\right]_{\alpha=1}, \quad N = 2m+2 \quad (12.49a)$$

$$\langle (E - \tfrac{1}{2}(\lambda_1 + \lambda_2)) \rangle = -(2)^{\frac{1}{2}}\sigma(m!)^{-1}\left[\left(-\frac{\partial}{\partial \alpha}\right)^m \alpha^{-1}\right.$$
$$\left.\times \left[\xi^{\frac{1}{2}} + \frac{1}{2}\left(\frac{\pi}{\alpha}\right)^{\frac{1}{2}}(\exp \alpha\xi)\mathrm{erfc}((\alpha\xi)^{\frac{1}{2}})\right]\right]_{\alpha=1}, \quad N = 2m+3 \quad (12.49b)$$

$$\langle (E - \tfrac{1}{2}(\lambda_1 + \lambda_2))^2 \rangle = \tfrac{1}{4}(\lambda_1 + \lambda_2)^2 + (N-1)\sigma^2 \quad (12.49c)$$

where $\xi = (8\sigma^2)^{-1}(\lambda_1 - \lambda_2)^2$ and erfc is the complementary error function (Abramowitz and Stegun 1965).

For large values of N, (12.49a) and (12.49b) become

$$\langle (E - \tfrac{1}{2}(\lambda_1 + \lambda_2)) \rangle = -\tfrac{1}{2}|\lambda_1 - \lambda_2|\{\tfrac{1}{2}(N-1)(\xi[\xi + \tfrac{1}{2}(N+1)])^{-\frac{1}{2}}$$
$$+ \xi^{\frac{1}{2}}[\xi + \tfrac{1}{2}(N-1)]^{-\frac{1}{2}}\}. \quad (12.50)$$

It is the quantity $\langle E \rangle \pm (\langle E^2 \rangle - \langle E \rangle^2)^{\frac{1}{2}}$ which indicates the possible correction which has to be applied to the lowest eigenvalue when the basis set is enlarged.

We now apply these ideas to the nucleus ^{16}O. The core is taken to be ^{12}C. For the ground state, $J^\pi = 0^+$ we have to consider four nucleons moving in the single-particle orbits $1p_{\frac{1}{2}}$ and $1d_{\frac{3}{2}}$. The single-particle energies are taken to be $\varepsilon_{p_{\frac{1}{2}}} = -4.95$ MeV and $\varepsilon_{d_{\frac{3}{2}}} = -3.3$ MeV. A 40-MeV Rosenfeld interaction is used to calculate the nuclear matrix elements. In the absence of configuration mixing all four nucleons would

12.4 PROBABILITY DENSITY FUNCTION

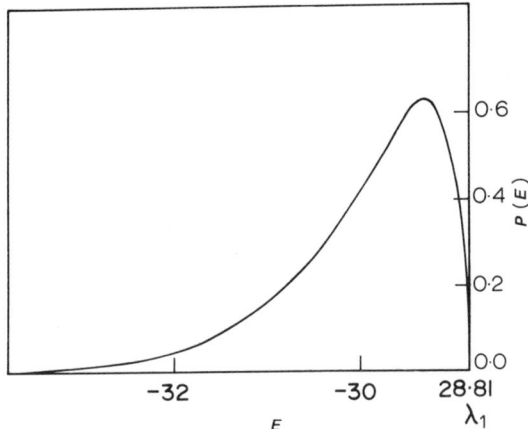

Fig. 12.3 Probability density function of the lowest eigenvalue for the ground state of the nucleus ^{16}O.

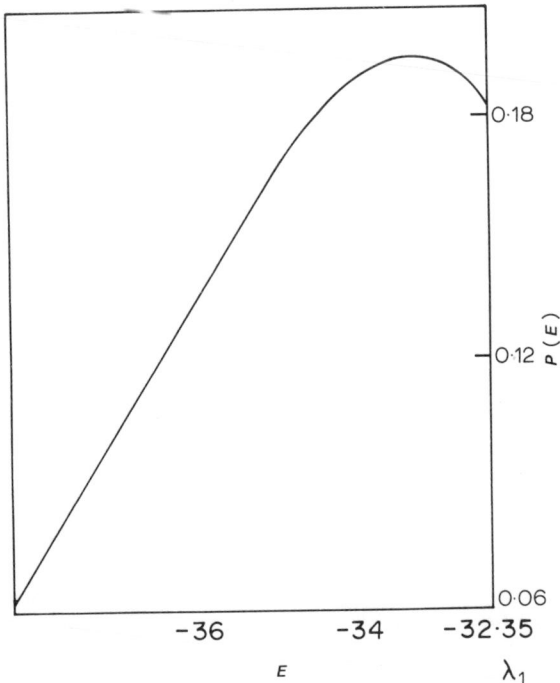

Fig. 12.4 Probability density function of the lowest eigenvalue for the first excited state of the nucleus ^{16}O.

have been in the 1p$_{\frac{1}{2}}$ level. The interaction mixes this configuration with other particle–hole configurations denoted by $(p_{\frac{1}{2}})^{4-2n}(d_{\frac{5}{2}})^{2n}$. The values of λ_1, λ_2, σ^2, and N for this case turn out to be

$$\lambda_1 = -32.35 \text{ MeV}, \qquad \lambda_2 = -26.17 \text{ MeV},$$
$$\sigma^2 = 16.93 \text{ (MeV)}^2, \qquad N = 3.$$

The plot of $P(E)$ versus E is shown in Fig. 12.3. Expressions (12.49) give the most probable range of the lowest eigenvalue as -33.17 to -37.72 MeV. This shows that the effect of configuration mixing on the lowest eigenvalue is of the order of a few MeV.

We also consider the excited state of the nucleus ^{16}O having $J^\pi = 3^-$. λ_1 is now the diagonal matrix element of the Hamiltonian with respect to the single-particle–hole configuration $(p_{\frac{1}{2}})^{-1}(d_{\frac{5}{2}})$. The interaction mixes this configuration with the three-particle–three-hole configuration $(p_{\frac{1}{2}})^{-3}(d_{\frac{5}{2}})^3$, giving

$$\lambda_1 = -28.81 \text{ MeV}, \qquad \lambda_2 = -17.31 \text{ MeV}, \qquad \sigma^2 = 4.88 \text{ (MeV)}^2, \qquad N = 4.$$

The probability density function $P(E)$ for this case is shown in Fig. 12.4. The effect of configuration mixing can be seen by calculating the mean and mean-square deviation using eqn (12.49). This calculation shows that because of configuration mixing the most probable range of the lowest eigenvalue ranges from -29.18 to -30.76 MeV. Thus, in the excited state case we find that the effect of configuration interaction is smaller than that for the ground state.

12.5 References

Abramowitz, M. and Stegun, F. A. (1965). *Handbook of mathematical functions*. Dover, New York.
Agassi, D., Gillet, V., and Lumbroso, A. (1969). *Nuclear Physics* **A130**, 129.
Anderson, T. W. (1958). *An introduction to multivariate statistical analysis*. John Wiley, New York.
Brown, G. E. and Jacob, G. (1963). *Nuclear Physics* **42**, 177.
Cohen, S. and Kurath, D. (1965). *Nuclear Physics* **73**, 1.
French, J. B. and Ratcliff, K. F. (1971). *Physical Review* **C13**, 94.
Kelson, I. and Levinson, C. A. (1964). *Physical Review* **134**, B269.
Kendall, M. G. (1959). *Advanced theory of statistics*. Charles-Griffin & Co., London.
Nesbet, R. K. (1955). *Proceedings of the Royal Society, London* **A230**, 312.
Porter, C. E. (1965). *Statistical theories of spectra. Fluctuation*. Academic Press, New York.
Rosenzweig, N. (1963). In *Brandeis University Summer Institute lectures in theoretical physics, 1962*, Vol. 3. Benjamin, New York.
Ullah, N. (1971). *Il Nuovo Cimento*, Ser. 11, **1A**, 49.

12.5 REFERENCES

Ullah, N. (1974). *Journal of Mathematical Physics* **15,** 880.
Ullah, N., Wong, S. S. M., and Trainor, L. E. H. (1970). *Canadian Journal of Physics* **48,** 842.
Wilkinson, D. H. and Mafethe, M. E. (1966). *Nuclear Physics* **85,** 97.
Wong, S. S. M. (1966). *Physics Letters* **20,** 188.
Wong, S. S. M. (1968). *Nuclear Physics* **120A,** 625.

Appendix A

EXPRESSIONS FOR THE FIRST FEW ORDERS OF ORDINARY PERTURBATION THEORY

Let H be the total Hamiltonian of the system which is written

$$H = H_0 + V \qquad (A.1)$$

where H_0 is the part which is exactly solvable and has eigenfunctions ϕ_n and eigenvalues ε_n.

Then we write the exact ground-state energy E_0 of H as

$$E_0 = \sum_{n=0}^{\infty} E_0^{(n)} \qquad (A.2)$$

where the various $E_0^{(n)}$ are given (Merzbacher 1961) by

$$E_0^{(0)} = \varepsilon_0, \qquad (A.3a)$$

$$E_0^{(1)} = \langle \phi_0 | V | \phi_0 \rangle, \qquad (A.3b)$$

$$E_0^{(2)} = \sum_{n \neq 0} \frac{|\langle \phi_0 | V | \phi_n \rangle|^2}{\varepsilon_0 - \varepsilon_n}, \qquad (A.3c)$$

$$E_0^{(3)} = \sum_{\substack{m \neq 0 \\ k \neq 0}} \frac{\langle \phi_0 | V | \phi_m \rangle \langle \phi_m | V | \phi_k \rangle \langle \phi_k | V | \phi_0 \rangle}{(\varepsilon_0 - \varepsilon_m)(\varepsilon_0 - \varepsilon_k)}$$

$$- V_{00} \sum_{m \neq 0} \frac{|\langle \phi_0 | V | \phi_m \rangle|^2}{(\varepsilon_0 - \varepsilon_m)^2}, \qquad (A.3d)$$

. . . .

Reference

Merzbacher, E. (1961). *Quantum mechanics*. John Wiley & Sons, New York.

SUBJECT INDEX

anticorrelation 39
anticorrelation identities 40
  examples 40
approximate evaluation of integrals 43
average of product of two scattering matrix elements 84
average reaction cross-section 66
average value of partial reaction cross-section using statistical collision matrix 105
  differential cross-section 108

band head 11, 12, 15
Bohr–Mottelson's collective Hamiltonian 28, 29

centroid 33
channel–channel correlation coefficient for complex boundary condition, absolute square of amplitudes 103
  partial widths 103
channel index 51, 93
  radius, surface 51, 93
Cohen–Kurath g.s. of ^{14}N 124
coherent phonon state 19, 26
collective vibrational Hamiltonian 18
complex amplitudes (elastic scattering) 96
complex orthogonal matrix 95
configuration interaction in ^{14}N nucleus using statistical methods 124
configuration interaction and statistical methods 115
correlation coefficient (rotational) 13
correlation coefficient of two eigenvalues 55, 60
correlation coefficient for two-electron system 36
  angular 36, 37
  in momentum space 36

diagonalization in truncated space 44
distribution of absolute square of complex amplitude of complex R matrix 100
distribution of eigenvalues for fixed diagonal matrix elements of Hamiltonian 127

distribution of lowest eigenvalue in two dimensions with fixed diagonal elements 129
  N dimensions 130
distribution of normalization constant 102
dominant component of g.s. of ^{14}N 124
dominant component of the many-body wave function 115

Ejiri's formula 25, 28
elastic scattering; R, S functions 52
energy correlation function 68, 87
energy levels of rotational nuclei 12
  rotation–vibration region 25, 27
  vibrational nuclei 19
ensemble
  circular 5
  Gaussian orthogonal 5, 79, 81
  Gaussian symplectic 5, 81
  Gaussian unitary 5, 81
  two-body random matrix 6
ergodic theorem 9
Ericson's fluctuations 67
exactly solvable Hamiltonian for two particles 37
expression for E_J in rotation–vibration region 28
expression for vibrational energy 20
expressions for first few orders of ordinary perturbation theory 134

Fourier transform of single eigenvalue for the three ensembles in two dimensions 83

graded vector 84
Grassmann integration 76
  examples 77
  expression for determinant as an integral over Grassmann variables 79

Haapakoski model 19, 26
Hartree–Fock and configuration interaction problem 117
Hartree–Fock wave functions 13

SUBJECT INDEX

Hauser–Feshbach expression 8, 67, 106, 113
Holmberg–Lipas model 15

level density 2
line element in the space of real symmetric matrices 54
   orthogonal matrix 53
Lipkin–Meshkov–Glick model 41
Lorentzian form 64, 87
low order moments for approximate evaluation of matrix elements 46

mean square deviation of single particle expectation value in the ground state 116
mean square fluctuation of cross-section using statistical collision matrix 105
   differential cross-section 110
   reaction cross-section 107
modification of perturbation theory using operator method 71
   linearization technique 73
moment of inertia of rotational nuclei 11
   energy dependence 15
   inverse 12
   semi-classical 16
   Skyrme 13
moment generating function for the expectation value of single particle operator 120
moments of complex amplitudes of complex R-matrix in two dimensions 96
moments (method of) 11, 15, 20, 110
multichannel reduced-width amplitude distribution 61
multi-level reduced-width amplitude distribution 57

nuclei ^{20}Ne, ^{28}Si 14
nucleus ^{152}Gd 29

overlap matrix element of rotation operator 48

partial reaction cross-section 8
partial width 33
particle–hole operators 18
partition function 16
perturbative statistical method 121
phonon creation operator 18
pole resonance form of scattering matrix 62

Porter–Thomas distribution 3, 57
position of resonance-width correlation 59
probability density function of partial total cross-section 110
probability density of Hamiltonian matrix elements 2
   eigenvalues 6
   expectation value of operator 4
   orthogonal matrix 4
   spacing 3
   widths 3
propagator using Lagrangian form of generating function 85

reduced matrix elements of quadrupole and dipole operators 124
reduced-width amplitude 51, 61
relation between average values of the parameters of statistical collision matrix 98
relation between R-matrix and S-matrix 51
relative and c.m. transformation 37
R-matrix 51
R-matrix for complex boundary condition 94
representation of determinant using angular momentum operators 88
resolvent (definition) 79
   average resolvent for large N 83
   average resolvent for two dimensions 80
   probability density of single eigenvalue using average resolvent 80
resonance correlations for complex boundary condition 104
Rosenfeld interaction 13, 23
RPA equation 18

Sanderson state 24
shell-model density 6
S-matrix 67
spacing distribution 55
spectra of Ce isotopes 26
square-root function 46
state density 2
statistical collision matrix 92
super multiplet symmetry of Wigner 35
symmetry mixing parameter 34

time-reversed state 93
traces of powers of H in two dimensions using angular momentum operators 90
transmission coefficient 64

SUBJECT INDEX

truncated space 126
two-point correlation function 86, 88

unitarity constraint on scattering
    function 58

vibrational parameters 20
    Peierls–Yoccoz formulation for
        vibrational parameters 21
    semi-classical expression 23
vibrational spectra of nucleus ^{114}Cd 19

volume element in the space of complex
    orthogonal matrices
    2×2 case 95
    $N \times N$ case 97

width fluctuation factor 106
width–width correlation
    elastic 59
    multichannel 63
Wigner's semi-circular distribution 55, 83
Wigner surmise 3
Wishart distribution 55

AUTHOR INDEX

Abramowitz, M. 45, 48, 49, 80, 81, 90, 129, 130, 132
Agassi, D. 116, 132
Anderson, T. W. 61, 69, 128, 132

Balian, R. 5, 76, 91
Banyard, K. E. 36, 42
Baranger, M. 17
Berezin, F. A. 77
Berthier, G. 42
Bhaduri, R. K. 16, 17, 32, 35
Biswas, S. N. 75
Bohr, A. 11, 13, 17, 25, 28, 29, 30
Bohr, N. 2, 50
Brink, D. M. 88, 89, 91, 110, 113
Brody, T. A. 3, 5, 6, 9
Brown, G. E. 116, 132

Chakraborty, M. 35
Chang, F. S. 33, 35
Cohen, S. 124, 132

Das Gupta, S. 16, 17
Datta, K. 75
Davidson, J. P. 29, 30
Del Re, C. 42
Devoigt, M. J. A. 11, 16, 17
Diamond, R. M. 31
Dirac, P. A. M. 7
Dudek, J. 17
Dyson, F. J. 5, 10, 81, 88, 105, 113

Edmonds, A. R. 88, 89, 91
Efetov, K. B. 76, 78, 91
Ejiri, H. 25, 28, 31
Ericson, T. 67, 69, 107, 110, 113

Feranchuk, I. D. 71, 75
Feshbach, H. 8, 67, 106, 110, 113
Flores, J. 9
Fock, V. 1, 36, 114
Franzini, P. 35
French, J. B. 6, 9, 33, 35, 130, 132

Gaudin, M. 55
Gillet, V. 116, 132
Glick, A. J. 41, 42
Gradshteyn, I. S. 22, 23, 24, 80, 91
Gupta, K. K. 19, 24

Haapakoski, P. 19, 24, 25, 26, 29, 30, 31
Hansen, E. R. 82, 91
Hartree, D. R. 1, 36, 114
Hauser, W. 8, 67, 106, 110, 113
Hill, D. L. 21, 24
Hioe, F. T. 71, 73, 75
Holmberg, P. 15, 17, 46, 49
Honkaranta, T. 24, 31
Hubbard, J. 86
Huizenga, J. R. 2, 10,

Inamura, T. 31
Ishihara, M. 31

Jacob, G. 116, 132

Katori, K. 31
Katriel, J. 40, 41, 42
Kelson, I. 114, 132
Kendall, M. G. 7, 8, 10, 120, 132
Kerman, A. K. 9, 10
King, F. W. 37, 38, 42
Komarov, L. I. 71, 75
Kota, V. K. B. 35
Kurath, D. 124, 132
Kutzelnigg, W. 36, 42

Lane, A. M. 50, 51, 52, 64, 69, 99, 108, 113
Levinson, C. A. 114, 132
Lipas, P. O. 15, 17, 24, 31, 46, 49
Lipkin, H. J. 41, 42
Lumbroso, A. 116, 132

Mafethe, M. E. 126, 133
Mahaux, C. 50, 69
Mehta, M. L. 5, 6, 7, 10, 52, 55, 69, 76, 91
Mello, P. A. 9

Merzbacher, E. 71, 75, Appendix A
Meshkov, N. 41, 42
Moldauer, P. A. 69, 87, 92, 97, 99, 108, 113
Monahan, J. E. 10
Montroll, E. W. 71, 73, 75
Moore, J. C. 36, 42
Moretto, L. C. 2, 10
Mottelson, B. R. 11, 13, 17, 25, 28, 29, 30

Nesbet, R. K. 74, 75, 114, 132
Ng, W. 12, 17
Nilsson, S. G. 13

Pandey, A. 9
Parikh, J. C. 35
Pastur, L. A. 83, 91
Peierls, R. E. 21, 24, 43, 45, 49
Porter, C. E. 3, 5, 6, 7, 10, 55, 57, 69, 94, 100, 113, 116, 132
Preston, M. A. 32, 35, 51, 70

Radicatti, L. A. 35
Ramond, P. 76, 78, 91
Ratcliffe, K. F. 130, 132
Ripka, G. 11, 13, 17, 48, 49
Rosenfeld, L. 13, 23, 35, 124, 125, 130
Rosenzweig, N. 9, 10, 116, 132
Rothstein, S. M. 37, 38, 42
Rowe, D. J. 18, 24, 73, 75
Ryzhik, I. M. 22, 23, 24, 80, 91

Sakai, M. 31
Sanderson, E. 20, 24
Sandhya Devi, K. R. 11, 15, 17, 58, 66, 70
Satchler, G. R. 88, 89, 91
Singh, V. 71, 75
Skyrme, T. H. R. 9, 10, 13, 17, 23
Slater, J. C. 1, 36, 114

Statonivitch, R. L. 86
Stegun, A. 45, 48, 49, 80, 81, 90, 129, 130, 132
Stephens, F. S. 31
Stephen, R. O. 110, 113
Szymanski, Z. 17

Tanner, N. W. 113
Thomas, R. G. 3, 7, 50, 51, 52, 57, 64, 69, 99, 100, 108, 113
Trainor, L. E. H. 12, 17, 133
Trudet, T. 9, 10

Ullah, N. 11, 15, 17, 19, 20, 24, 27, 29, 31, 33, 34, 35, 38, 40, 41, 42, 56, 58, 60, 61, 70, 71, 75, 94, 113, 117
Urbano, J. H. 43, 45, 49

Verbaarschat, J. J. M. 7, 9, 10, 69, 70, 76, 83, 85, 86, 87, 88, 91

Ward, D. 25, 31
Warke, C. S. 58, 60, 70
Watson, C. N. 113
Weidenmüller, H. A. 10, 50, 69, 70, 83, 91
Wheeler, J. A. 21, 24
Whittaker, E. T. 113
Wigner, E. P. 2, 3, 4, 10, 35, 50, 52, 70, 76, 93, 95, 113
Wilkinson, D. H. 126, 133
Wishart, J. 55
Wong, S. S. M. 6, 9, 10, 33, 35, 118, 124, 133

Yoccoz, J. 21, 24

Zinn-Justin, J. 76, 91
Zirnbauer, M. R. 10, 70, 83, 88, 91

RAYMOND H. FOGLER LIBRARY